AF255293

SECOND GEOLOGICAL SURVEY OF PENNSYLVANIA:
1874.

REPORT OF PROGRESS

IN THE

VENANGO COUNTY DISTRICT.

BY JOHN F. CARLL.

OBSERVATIONS ON THE GEOLOGY AROUND WARREN.

BY F. A. RANDALL.

NOTE ON THE COMPARATIVE GEOLOGY OF NORTH-EASTERN OHIO AND NORTH-WESTERN PENNSYLVANIA, AND WESTERN NEW YORK.

BY J. P. LESLEY.

HARRISBURG:
PUBLISHED BY THE BOARD OF COMMISSIONERS
FOR THE SECOND GEOLOGICAL SURVEY.
1875.

BOARD OF COMMISSIONERS.

His Excellency, JOHN F. HARTRANFT, *Governor*,
and *ex-officio* President of the Board, Harrisburg.

ARIO PARDEE,	Hazleton.
WILLIAM A. INGHAM,	Philadelphia.
HENRY S. ECKERT,	Reading.
HENRY M'CORMICK,	Harrisburg.
JAMES MACFARLANE,	Towanda.
JOHN B. PEARSE,	Philadelphia.
ROBERT B. WILSON, M. D.,	Clearfield.
Hon. DANIEL J. MORRELL,	Johnstown.
HENRY W. OLIVER,	Pittsburg.
SAMUEL Q. BROWN,	Pleasantville.

SECRETARY OF THE BOARD

JOHN B. PEARSE,	Philadelphia.

STATE GEOLOGIST

PETER LESLEY,	Philadelphia.

PLEASANTVILLE, *Feb.* 15, 1875.

Prof. J. P. LESLEY, *State Geologist:*

DEAR SIR:—I have the honor to submit herewith a brief report of the progress of the work under my charge in the Oil District of Venango, in connection with the Second Geological Survey of Pennsylvania, during the season of 1874.

Very respectfully,

JOHN F. CARLL.

TABLE OF CONTENTS

OF THE SEVERAL CHAPTERS.

I.

IN THE VENANGO OIL DISTRICT, 1874.

BY JOHN F. CARLL, ASSISTANT GEOLOGIST.

CHAPTER I.

Outline of Work.   Datum of Levels.   Contouring and Maps.

Field work, in the Venango district, was commenced on the 16th of July, and continued, as the weather permitted, until about the 10th of December.

Base lines have been run, elevations established, specimen fossils collected and facts noted, not so much with a view of completing any particular portion of the survey, as of furnishing the outlines for its more rapid execution during the coming year.

As our party consisted only of Mr. F. A. Hatch and myself, and the unusually fine weather of the past season permitted so much time to be occupied in the field, where the presence of both was always required, little opportunity was afforded for office work.   Consequently, as this report of progress was called for early in the winter, the collation of facts and the systematic study of results in detail must, in a large degree, be postponed to the coming year ; and much of the work, actually done, can only appear by implication in the general statement.

Tide Level datum.

The first thing required in this district, where so much depends on accurate levels, was a base above tide to work from. This was obtained from data furnished by the Allegheny Valley and Oil Creek railroads.   Taking these in preference to others as being the most central to the field of examination, three lines of levels have been run from " Ennis Hill," Pleasantville, the highest point in Venango county, to intersect these railroads at

Rouseville, Tionesta and Tidioute; and the result (which did not vary a foot on either line) places the summit of Ennis hill at an elevation of 1,713 feet above tide. This datum, however, may be held subject to future correction; as some discrepancy appears after checking with points given on the Philadelphia and Erie, and Dunkirk and Allegheny Valley railroads. But this will not affect the harmony of our calculations, since our elevations are all referred to the tide level, and used only relatively to each other.

Unfinished Work.   Contouring and Map.

Owing to the late date in the season when the instruments for contouring and measurements arrived, little has been accomplished under this head. We have not had opportunity to complete the bearings and measurements of all our base lines. Consequently the map appended to this report, which is intended to show the location of wells, and of the oil-producing centres spoken of in this paper and not given on ordinary maps, was necessarily constructed in part from other maps, and cannot claim for itself to be correct in every particular.

To the above evidences of the unfinished state of our work may be added the fact that levels were taken to scores of wells of which no records could as yet be obtained, and these levels will come into use for our special purpose when the records are recovered.

These statements are sufficient to show why no finished report should be expected at this time. The field is large and comparatively new, so far as precise and accurate geological work is concerned. Only a general knowledge of the oil bearing strata is, at present, claimed by geologists, and it would be unwise, after only a few months of study in this really first attempt at an exhaustive investigation, to offer anything but a brief report of progress.

Topography.

Of the topographical features and facial characteristics of this district, its water sheds and the directions of its streams, no detailed description is necessary. All this will be better exhibited by a future contour-line map, for which materials have been collected. Our surveys in 1874 have been confined to a small area; the general features are well understood, being common to the whole county lying between the New York and Ohio State lines; while the special features of Venango county can only be made clear after the instrumental work of the survey shall have been pushed forward over a larger extent of territory.*

Our principal work was directed to the study of the oil bearing strata, for which the term " sands " is universally used; and we shall proceed at once to a consideration of the facts thus far developed by the survey.

In speaking of localities, sand-rocks and matters relating to oil wells, we shall use the terms in vogue here and generally understood by oil operators, although they may sometimes have an unscientific sound and in a few instances seem almost absurd.

---

* [A general statement of the topography and geology of the region south of Lake Erie, intended to supply what Mr. Carll judiciously refrains from inserting in his report, and to give to citizens of other districts of the State some idea of what is so familiar to the citizens of Western Pennsylvania, will be found in this volume. See Article II. J. P. L.]

CHAPTER II.

The Genesis of Oil.   Ambiguity of terms.   Discovery of the Oil Sands.   Belts versus Streams.   Barren Oil Measures.

Concerning the origin of oil, so little definitely is known that it would be presumptuous to enter upon a discussion of that subject. Many theories have been advanced, but none of them meet the full requirements of the facts developed by the drill. The coarser and more porous the stratum of pebbles, as a general rule, the larger may the well be expected to prove. But what there is in this sand to generate the oil no one has been able to demonstrate. In all the drillings brought up by the sand-pump and in all examinations of outcrops of oil sands, I am not aware that anything has been detected which can be supposed to have originated the oil in the rock itself. Whence then does it come? Certainly not from *above;* for being lighter than water and generally resting under a heavy pressure of gas, its tendency is always upward; a fact plainly exhibited by oil springs and flowing wells. If, generated and developed *as oil* at any depth *below* the oil bearing sands, it has traveled upward through fissures in the rocks to the sands above, why has it not happened that some one, out of the large number of wells that have been sunk hundreds of feet below the third oil sand, has struck these upward leads and produced oil? Indeed, if it travels at all *as oil* from sand to sand why are not these leads *frequently* tapped between sands?

But a few cases have been reported where the oil was claimed to have been obtained between the sands, and in these instances we believe a careful investigation into the facts would have led to a satisfactory explanation of these apparent exceptions to a general rule.

The history of all the deep wells drilled in the oil regions may be written in one word—"failures." Wells have been put down at Titusville, on Oil Creek, on French creek, along the Allegheny river between Franklin and Tidioute, at Tidioute and above, on the Brokenstraw, Little Brokenstraw and Hosmer run, to depths varying from 200 feet to 1,900 feet below the horizon of the third-sand of Oil Creek, and whatever may have been the claims of the drillers as to sands and " good shows," the fact stands uncontradicted, they were all unproductive. In most of them it was admitted that no well-defined sand, like the oil bearing sand, was found below the plane of the regular third-sand of Oil Creek.*

We leave these questions unanswered, having merely adverted to the subject to show the importance of the sand-rocks in the petroleum measures, whatever their functions may be, and the consequent necessity of devoting the principal part of this report to their consideration.

Ambiguity of Terms.

It may be well in the outset to state that a great deal of ambiguity has been permitted in speaking and writing of the oils and oil rocks ot this district. No strict classification of either has been adhered to. We have heard of oil sands numbered from one to six, and of oils—amber, green and black, with gravities ranging from 30° to 55°—without inquiry into the relative positions of these sands, and the reasons for this variety in the oils; or any attempt to refer certain oils to certain sands.

In like manner we have been told of "Oil Belts" lined out over black oil territory and green—light gravity oil and heavy— promiscuously.

That such confusion of terms may be avoided, an essay will be made in this report to arrange these sands in regular order, that they may be studied more understandingly.

---

* Before this report went through the press, that is, in April, 1875, much excitement was caused by the sudden and unexpected flow of oil from Mr. Beattie's gas well at Warren. Whatever may be the future history of this well, it will not impair, in any material feature, the correctness of Mr. Carll's statements given on this and the following pages; for the geological horizon at which the Beattie well gets its oil, lies at a depth of at least 600 feet below the base of his petroleum measures. The case is also, as yet, an isolated instance; or belongs to a totally different and, as yet, unstudied oil region lying in M'Kean, Elk and Potter counties. Even if a productive oil district be found to lie around and east of Warren, the fact remains all the same true, that the *productive oil series* of Venango, Armstrong and Butler is a well-defined group, restricted to three principal gravel-sand strata and their subdivisions, as described by Mr. Carll.   J. P. L.

First Artesian Wells: Discovery of Sands.

In the first oil development by Artesian wells, nothing was known about the sands. Wells were drilled until indications of oil appeared, without regard to the character of the strata pierced. But experience soon proved the sand rocks to be its source, and then commenced deeper drilling for other sands, which, in the valley of Oil Creek, resulted in the discovery and classification of "three sands"—these being all the oil bearing sands found in that locality, even after several wells had been sunk much deeper in quest of others.

In the progress of development locations for wells were selected on higher ground. The drill passed now through four or five other and higher definite sand-rocks before reaching the geological horizon of the First Sand of Oil Creek. When this fact was made clear it became customary among drillers to throw out these *upper* sands from their well-records. They were called the "Mountain Sands," and were also numbered 1, 2, 3, &c. The drillers commenced their count of the oil-rocks with that one which they found at the depth at which they supposed the First Sand of Oil Creek to lie. But in doing so, many errors occurred, resulting from a want of accurate information: first, as to the surface elevation of the wells drilled on high ground; and second, as to the dip of the oil bearing strata, which materially affected the comparison of elevations, even when these were accurately known.

A third source of error may be found in the fact, that a thick stratum of sand lying single and solid in one place is often split into two, or, in other words, is represented by an equivalent of two sands with shales intervening, in another place, perhaps only a short distance from the first. This will be illustrated in the progress of this report.

Belts versus Streams.

For several years after the discovery of oil the drilling of wells was almost exclusively confined to the "flats" bordering the principal streams. The impression prevailed that there was some connection, some parallelism between the streams on the surface and the "oil-veins" beneath. But the many failures to strike oil along the streams gradually led to locations on higher ground, and upon lines between good wells. This method has been pursued so long and so thoroughly, that we can now affirm that the drill has traced the great oil-leads of the country from point to point *regardless of any and all topographical features of the surface.*

It seems to have been eminently proper that the initiatory steps of this survey should be taken in this part of the oil field where the large number of wells, in all directions, have so clearly defined the trend of the oil bearing rocks, and where the history of so many unsuccessful ventures will also furnish much valuable information.

Mountain-sand Group or Barren Oil-Measures.

We use the word " belt," not as employed by some to designate a narrow, continuous line of sand-rock, which may be uneringly traced for miles, with an instrument, on a certain degree of the compass circle; but only as a convenient term for expressing the general trend of the oil bearing rocks from point to point; even although interrupted by " dry" and unproductive intervals.

The base line run from Pleasantville to Tidioute—from the commencement of the Colorado district to the Allegheny river—passes through what has been one of the best and most continuous oil producing belts of the region. Along, and contiguous to, this line, and to the north of it, the deeply eroded valleys of Pine creek and Dennis run expose the basset edges of the whole series of slightly inclined rocks (uplifted towards the north) underlying the Great Conglomerate (No. XII, the base of the productive coal measures) to a (geological) depth of 850 feet, bringing us down to within about 100 feet of the third or lowest oil-bearing sand.

Not having had a proper opportunity for running a line of levels from any of the unmistakable outcrops of the coal measure Conglomerate No. XII, to connect it with our base on Ennis hill, we have been compelled to assign it its position there from collateral evidences which appear later in this report; and while feeling very little doubt about its location, we admit, of course, the possibility of some unknown error which may call for a local re-adjustment hereafter.

This exposure, (along Pine creek and Dennis run,) taken in connection with the well records along the route, enables us to form a tolerably correct idea of the stratification of the rocks to that depth. The whole series is found to consist of bands of sandstones and conglomerates, and sandy and muddy shales and slates, varying locally in character, composition and relative order, when studied in detail, but as a whole lying one above another in nearly horizontal parallel planes. The local variability of stratification is particularly noticeable (at least in the south-eastern part of the district) in the strata next beneath the conglomerate No. XII, and to a relative depth of from 600 to 650 feet. These strata have never produced oil in Venango county. We may, therefore, call them the Barren Oil-measures of Venango; or the Mountain-sand Group.

Oil-sand Group, or Petroleum Measures.

Beneath the division of mountain-sands another series, with a thickness of from 350 to 400 feet, and similar to the above in structure, but rather more regular in stratification, will include the three sands of Oil Creek; and, as we believe it can be shown that no oil has ever been obtained in the district except from rocks of this series, it may properly be called the Petroleum Measures of Venango, or Division of the Three Sands.

Some of the first wells drilled evidently obtained their oil above the First Sand—and the old Oil Pits of French and Oil Creeks and Hosmer Run were above it also. But the oil, without doubt, came really from this First Sand, its close proximity to the surface in these places having admitted of the percolation of surface water into its crevices, which, by hydraulic pressure, forced the oil upward.

It is a noticeable fact, that any first sand below the surface is generally full of water veins, whether it be an Oil-bearing or a Mountain sand. If the Oil sands lie deep they seldom (especially in new territory before the water is let down by the drill) contain much water.

In the shallow wells at Tidioute, along the Allegheny River, on French and some parts of Oil Creek, considerable water was always pumped with the oil. But in the deep wells at Pleasantville there was not found at first one per cent. of water, and that, being salt, must have come commonly from the second sand.   As the oil was exhausted, the water increased.

We can learn of no reliable instance of oil being obtained above the First Sand where it might not be attributed to these results of its nearness to the surface.

As to the failures to obtain oil below the Third Sand see page 9.

We, however, in this report, carefully restrict the use of the above terms—Mountain-sand and Oil-sand Measures; Barren Oil and Petroleum groups—to Venango county, and have no intention to apply them broadly to the whole Oil Region of Western Pennsylvania.   They are convenient, and in a geological sense necessary and proper.

CHAPTER III.

The Petroleum Measures: Oil-Sand-Group.

A comparison of records of wells on Oil Creek, where the three leading sands of the Petroleum Measures lie with considerable regularity, both as to their thickness and the intervening distances between them, results in an average record about as follows:

First Sand, 40′ feet thick.
Interval - - 105′.
Second Sand, 25′ feet thick. } 315′.
Interval - - 110′.
Third Sand, 35′ feet thick.

In addition to these three regular sands there is found in many of the wells a fine-grained, muddy, gray-sand, known among drillers as the "Stray Third." This lies from 15 to 20 feet above the regular Third, and is from 12 to 25 feet thick. In some localities this rock assumes a pebbly character and produces oil, which is always darker than the Third Sand oil; sometimes being nearly black.

At different points on Oil Creek—at East Shamburg and other places—wells in close proximity to each other have produced, some of them black oil, some green, and some a mixture of both.

The "black oil" of the Pleasantville district has all been derived from the "Stray Third," which, in this district, is universally called the Fourth, or "Black Oil Sand."* But here the character and composition of the two sands (Third and Stray) are reversed. The Stray is a coarse pebble or conglomerate; the Third, a fine, micaceous, muddy, gray sand, only 15′ to 20′ feet in thickness; but always showing traces of green oil, and sometimes furnishing an abundance of gas.

We believe it can be shown also, that Pithole, Cashup and Fagundus, although producing an oil of lighter color than Pleasantville, drew their supply from the same Stray Sand, and the proof will be offered further on.

---

* The Second Oil Sand being here divided into two, as will be shown further on.

A noticeable peculiarity of these two sands (Third and Stray) is, that on the north-western outline of the oil field, where the Third shows itself in greatest force, the Stray is seldom an oil producing rock. As we proceed south-eastward the Stray begins to get its pebbly constitution and to yield oil over broader areas than the Third; the latter becoming more fine and compact, and gradually thinning away.

A marked difference will be noted also on comparison of specimens of the two sands. In the oil-producing Stray the pebbles are of a yellowish brown color, and in shape generally spheroidal. In the Third the pebbles are white, often brilliant, and in shape lenticular. These distinguishing characteristics, we believe, hold good universally.

On the north-westerly line above mentioned the Second Sand lies in a massive stratum, 30′ feet, or more, in thickness. Towards the south-east—as in a part of the Pleasantville district, at Bean Farm, Pithole, Cashup and Fagundus—it is split into two well defined sands, with from 15 to 30 feet of slates or shales intervening.

It is this that has given rise to the erroneous appellation of Fourth Sand oil at Pleasantville. The drillers began to number rightly on the First; called the split (Second) sand next below it Second and Third; and then called the Stray the Fourth. This of course made the Third Sand of the Oil Creek wells, which was still lower, Fifth in the Series.

In some localities they went still farther in their zeal to prove their territory better than Oil Creek, by showing a greater number of sands.* Finding the Stray and Third in *three* divisions instead of two, they announced at once the discovery of a Sixth sand.

The First sand, as far as we have examined it, appears to lie with more uniformity than the Second; but further investigation may show changes of character and of level similar to the others.

Little oil has been produced from the First and Second sands in the particular field under review. Their best development, as oil-bearing rocks, is along the Allegheny River from West Hickory to the Cochran farm, and on French Creek and Two Mile

---

* It must be understood that the fashion became general to estimate the value of oil lands according to the number of "Sands" passed through by the wells.

Run, near Franklin, to which our detailed survey of 1874 did not reach. We speak of them above as they are found on the green oil range, and without a closer knowledge of the peculiar structural differences which they may be found to exhibit in the places above named, on the Allegheny river and French Creek.

Assuming then that all the oil from this county has been deduced from the " Group of the Three Oil Sands," consisting of the First, Second, Stray and Third, with their intervening slates, shales and mud rocks, and that the trend of the oil-producing belt is marked by no surface indications to point out its direction or drift, we will proceed, on the principle of a general parallelism of strata, to trace the sands by means of the levels run, combined with the records of wells, through some of the main oil centres of the district; with a view of ascertaining the direction of the dip of the series and the fall, in feet, per mile.

Trend of the Upper Oil Belt.

The Venango Petroleum District, or " Upper Oil Belt," as it is now generally called, in contradistinction to the Butler county district, may be said to commence a short distance east of Tidioute. From thence southwestward it is marked by an almost unbroken band of wells, through Dennis Run, Triumph, the Clapp farms, New London, the Ware farm and Colorado, a distance of about nine (9) miles.

Between this, its south-west end, and the commencement of the Shamburg district, near the National wells, no paying Third Sand wells are found, except, perhaps, within a limited area on the Benedict Farm, west of Enterprise, the exact geological relations of which, to the Colorado " lead," has not been fully determined.

Beneath this unproductive district the Third Sand is found in all the wells drilled, having a thickness of from 30 to 45 feet, but apparently too fine-grained and closely compacted with mud to produce oil.

Between Shamburg and Petroleum Centre, on Oil Creek, occurs another unproductive interval. But from Petroleum Centre the oil belt has been traced with considerable continuity, crossing the Allegheny River at Reno, again at Foster's, and terminating at Scrubgrass.

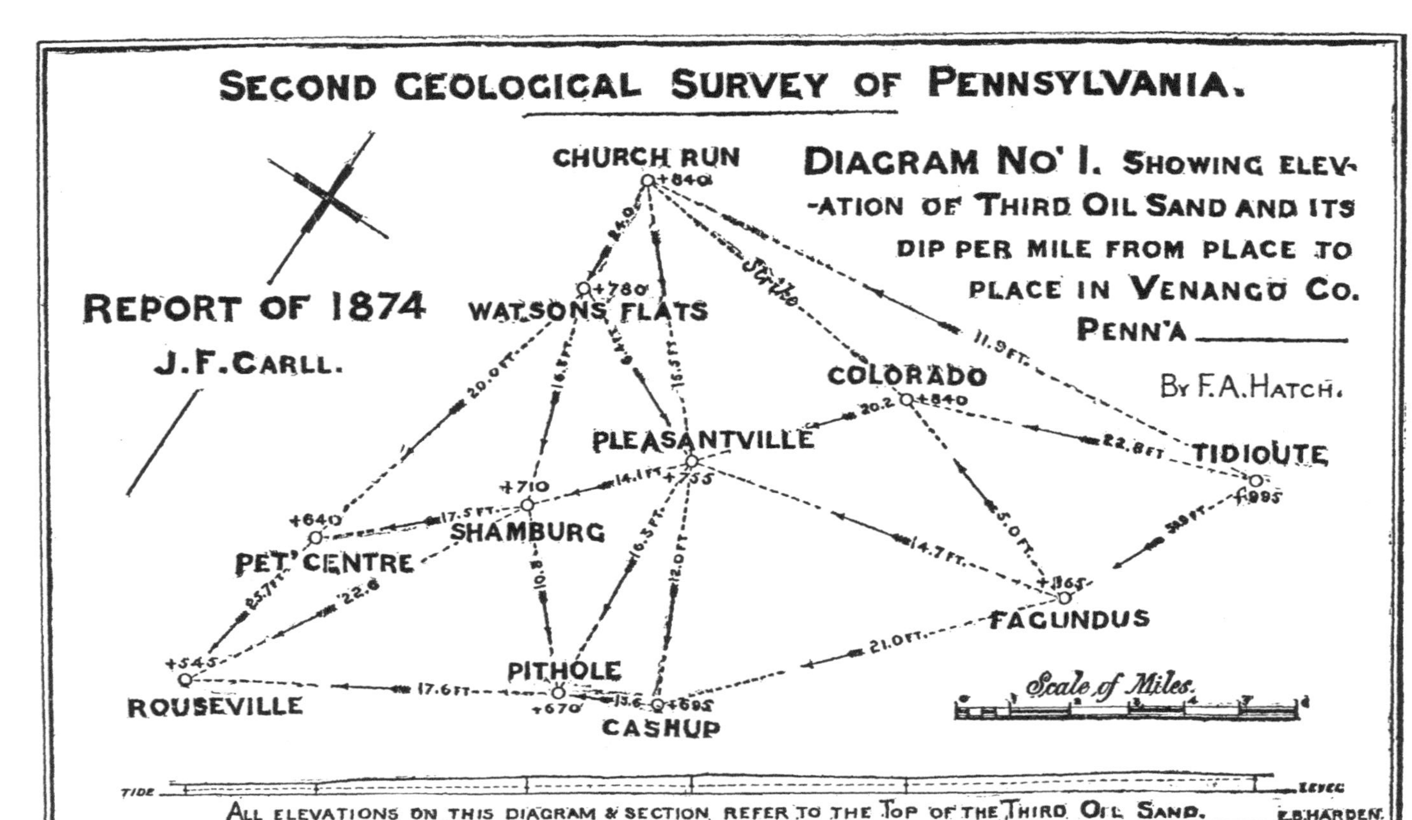

SECOND GEOLOGICAL SURVEY OF PENNSYLVANIA.
REPORT OF 1874
J.F. CARLL.
CHURCH RUN
DIAGRAM No' I. SHOWING ELEV-
-ATION OF THIRD OIL SAND AND ITS
DIP PER MILE FROM PLACE TO
PLACE IN VENANGO CO.
PENN'A
By F.A. HATCH.
WATSONS FLATS
COLORADO
PLEASANTVILLE
TIDIOUTE
PET' CENTRE
SHAMBURG
FAGUNDUS
ROUSEVILLE
PITHOLE
CASHUP
Strike
Scale of Miles.
TIDE
LEVEL
ALL ELEVATIONS ON THIS DIAGRAM & SECTION REFER TO THE TOP OF THE THIRD OIL SAND.
E.B. HARDEN.

As the survey has not yet been extended beyond Oil Creek, the lower or southern portion of the "belt" will not be discussed.

This line of development, it will be noted, leaves Tidioute in a direction of about S. 80° W., gradually sweeping around toward the south, and ending with a bearing of only about S. 20° W.

The belt above described, it should be understood, is the green oil or Third Sand belt. It appears to be much narrower and more sharply defined than others. At many places a distance from the centre-line towards the north or towards the south of merely a few rods suffices to guarantee a " dry hole."

Elevation above Tide and Dip of Oil Sands.

From levels taken along the surface line above described, combined with such records of wells as were obtained, the elevation of the top of the Third sand in the several localities named, is ascertained to be as follows:

At Tidioute	995 feet above tide.
" Colorado	840 " " "
" Pleasantville	755 " " "
" Shamburg	710 " " "
" Petroleum Centre	640 " " "
" Rouseville	545 " " "

Distance from Rouseville to Tidioute, 20.7 miles.
Difference in elevations,   -   -   -   - 450 feet.
Dip per mile,   -   -   -   -   -   - 21.7 "

That a clearer understanding may be had of this dip, which is subject to local variations, resulting in curved lines both for the dip and the strike, we append the following diagrams:

No. 1. Giving the dip from place to place, calculated from the elevations of the Third sand above tide, and the distances between points. It shows the relative positions of the places above named, with the elevation above tide of the top plane of their respective Third Sands. Titusville (Watson Flats) on the north, and Pithole, Cashup and Fagundus on the south, have been added to the chart; the dip of the rock *per mile* from place to place being stated on the joining lines in feet.

The exact relation of Titusville and Church Run to the centre line of development we have not as yet the means of fully determining. Considerable discrepancy appears in the well records;

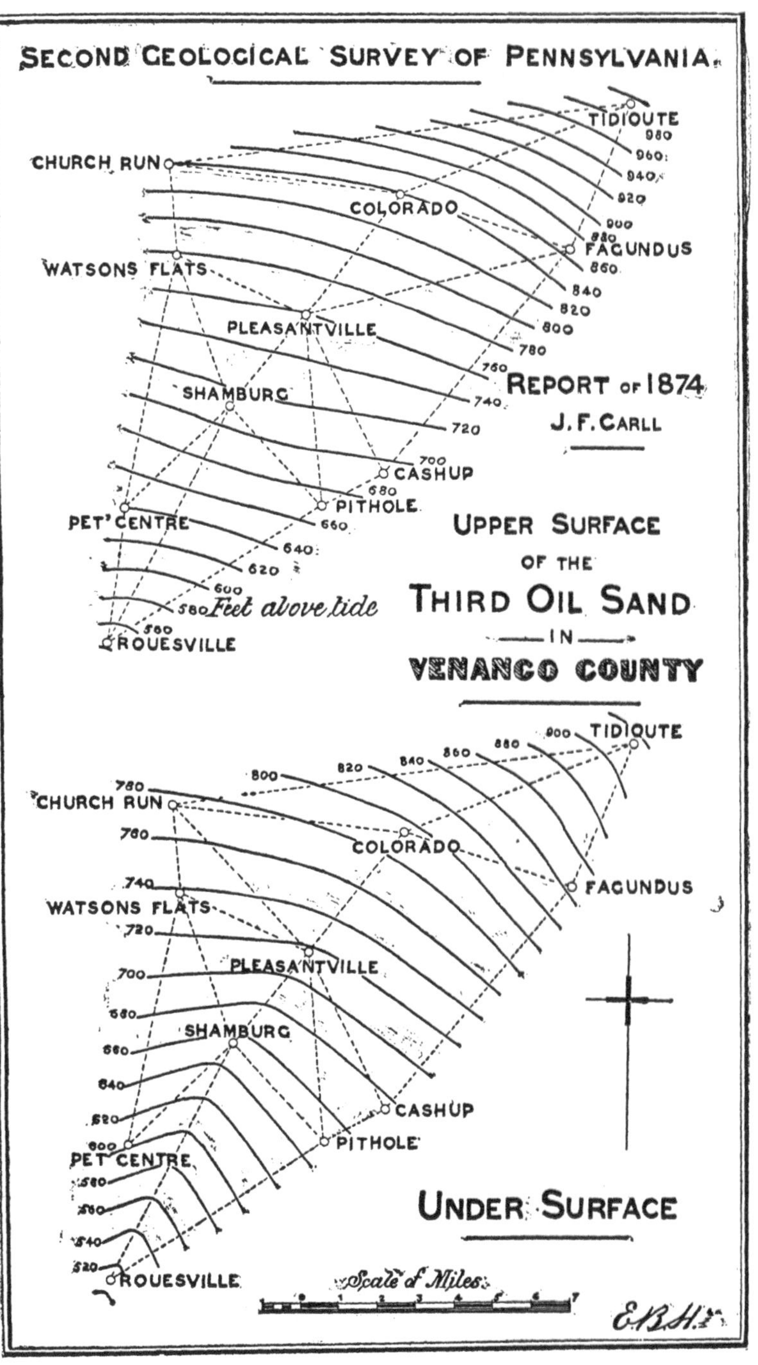

SECOND GEOLOGICAL SURVEY OF PENNSYLVANIA.
TIDIOUTE
980
960
940
920
CHURCH RUN
900
COLORADO
880
FAGUNDUS
860
WATSONS FLATS
840
820
800
PLEASANTVILLE
780
760
REPORT OF 1874
740
SHAMBURG
J. F. CARLL
720
700
CASHUP
680
660
PITHOLE
PET' CENTRE
640
620
UPPER SURFACE
600
580 Feet above tide
OF THE
560
ROUESVILLE
THIRD OIL SAND
IN
VENANGO COUNTY
900 TIDIOUTE
880
860
840
820
800
780
CHURCH RUN
760
COLORADO
740
FAGUNDUS
WATSONS FLATS
720
700
PLEASANTVILLE
680
SHAMBURG
660
640
620
CASHUP
600
PITHOLE
PET CENTRE
580
560
UNDER SURFACE
540
520
ROUESVILLE
Scale of Miles
E.B.H.

particularly in those bored on Watson Flats. The oil rock appears to belong to the same plane as the Third sand of the Lower Oil Creek; the Second being absent; and the First about in its proper place.

The oil-producing rocks of Pithole, Cashup and Fagundus would seem, from a study of the diagrams and a comparison of well records, to belong, as before intimated, to the horizon of the Stray.

This view of the case is strengthened by the fact that another sand is found here (at Pithole, Cashup and Fagundus) underlying the oil-producing rock, the same as at Pleasantville. And this other or lower sand occupies a position relative to tide level agreeing with the *general* dip of the Third sand. Were we compelled to regard the oil-producing rock of these places as identical with the Oil Creek Third, we should, in that case, have a *rise* in the rock from Shamburg to Pithole and Cashup, and from Colorado to Fagundus; reversing the general dip, and entirely destroying its uniformity; in other words, indicating a want of parallelism in the strata not noticeable elsewhere.—[A Diagram prepared to show this is omitted.]

In the Pleasantville district we have convincing proof that the oil came from the Stray. More than fifty wells have been sunk deeper, all of them finding from 15 to 20 feet of Third sand in place beneath, but without oil, although in many cases furnishing a large supply of gas.

In the neighborhood of the National wells and at East Shamburg—where the pebbly stratum of the Stray crosses, or overlaps, the pebbly stratum of the Third—large wells were obtained in the Third sand, after the Stray had been exhausted; and we are confident, that, as the survey advances and more well-records on the cross section of the field are procured, there will be no difficulty in tracing the Stray, step by step, from East Shamburg and Pleasantville, to Pithole and Cashup. Why this Stray thickens and becomes a better oil-bearing rock toward the south-east, while the underlying Third thins away into thin bedded, laminated, muddy sands, is a problem for future investigation.

Diagram No. 1 discloses another fact. While the *general dip* of the strata from Tidioute to Petroleum Centre is in a south-westerly direction it is not a perfect plane, but a "warped sur-

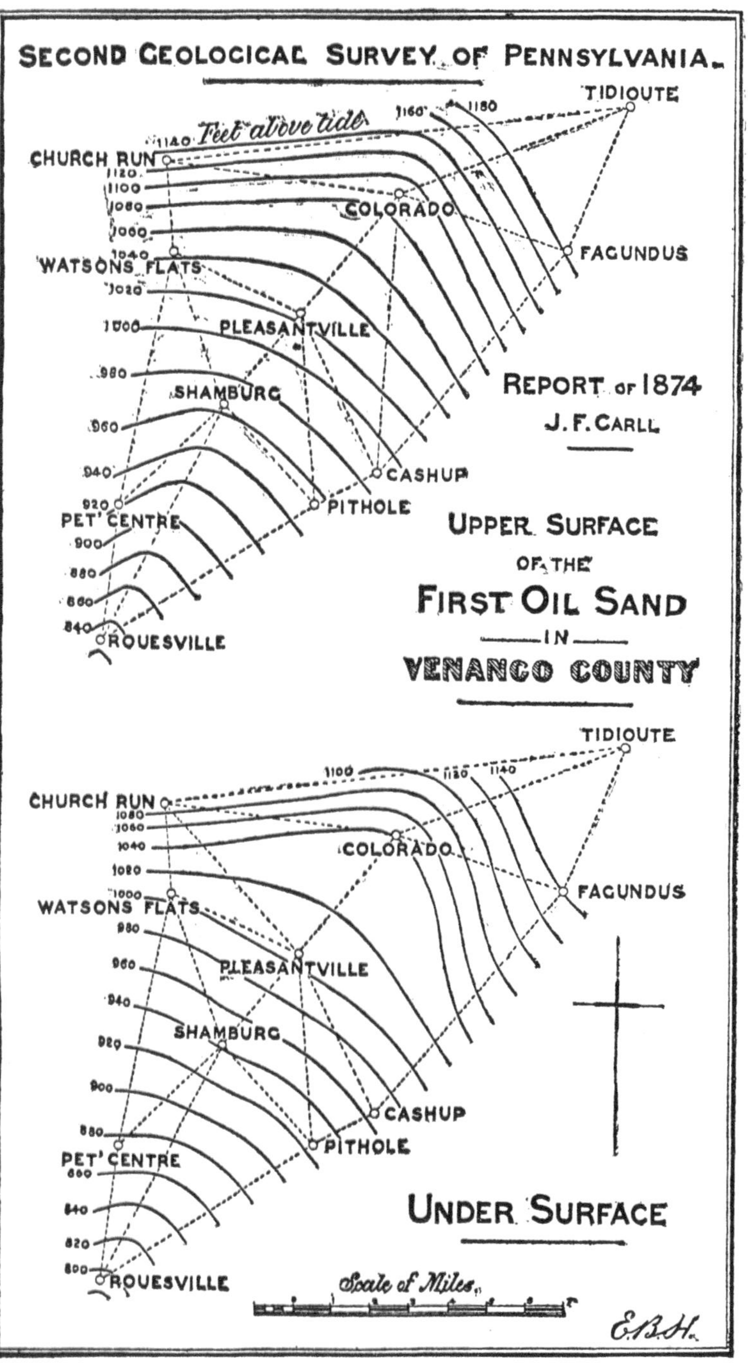

SECOND GEOLOGICAL SURVEY OF PENNSYLVANIA.
Feet above tide
TIDIOUTE
1160
1180
CHURCH RUN
1140
1120
1100
1080
1060
COLORADO
1040
WATSONS FLATS
1020
FAGUNDUS
1000
PLEASANTVILLE
980
SHAMBURG
960
940
CASHUP
920
PITHOLE
PET' CENTRE
900
880
860
840
ROUESVILLE
REPORT OF 1874
J. F. CARLL
UPPER SURFACE
OF THE
FIRST OIL SAND
IN
VENANGO COUNTY
TIDIOUTE
1100
1120
1140
CHURCH RUN
1080
1060
1040
COLORADO
1020
1000
WATSONS FLATS
980
FAGUNDUS
960
PLEASANTVILLE
940
SHAMBURG
920
900
CASHUP
880
PITHOLE
PET' CENTRE
860
840
820
800
ROUESVILLE
UNDER SURFACE
Scale of Miles
E. B. H.

face." The points under Colorado, Pleasantville and Shamburg lie somewhat (and in different degrees) *below* a plane passing through Tidioute and Petroleum Centre. They lie also below a plane passing through Church Run and Pithole. This wave-like feature in the formation is what might be expected as a result of causes producing the general uplift towards the north.

A part of the unevenness and want of resemblance shown by the contour lines (of dead level, or strike) of the upper and under sides is evidently due to the varying thickness of the Third sand, the distance between its upper and under surfaces being sometimes 20 and sometimes 70 feet. But it is not yet ascertained whether the sands thicken and thin from a regular plane below upwards, or from a regular plane above downwards.

Diagrams Nos. 2 and 3, on plate 2, page 19, are intended to show the warped upper and under surfaces of the Third Oil Sand formation, each contour-line representing a dead level along the face of the rock at 520, at 540, at 560, &c., feet above tide. The lines therefore represent the *varying local strike;* and, of course, the varying local *dip* will be at right angles to the same, downwards, in directions converging towards Rouseville. In the country next west of the district thus represented these contour-lines, if continued, would be seen to swing round into their regular south-west course, parallel to the shore of Lake Erie.

Diagrams Nos. 4 and 5, on plate 3, page 21, make the same exhibit of *strike* and *dip* for both the top and bottom surfaces of the First Oil Sand formation. The excessive steepness of its bottom surface *east and north of Colorado,* followed by an excessive flatness *south-west of Colorado,* is explained—(1.) by the thickening of the formation from Fagundus (Section No. 13) and Rooker Farm (No. 12) to Colorado (No. 7; See plate 5, page 27;)—(2.) by the drop of the whole rock, or, in other words, its approach towards the Third Sand;—(3.) by the reversal of both the above, viz: by the thinning of the rock, and by the increase of its distance from the Third Sand, between Colorado and Pleasantville.

If, however, the one First Sand at Colorado be considered the equivalent of two rocks at Fagundus (Sec. 13), and if the lower of the two have been overlooked by the drillers at Rooker Farm (Sec. 12,) then the system of contours in Diagram 5 (plate 3, p. 20) must be re-drawn.

Sections constructed from Records of Oil Wells.

A group of sections drawn from well-records and typical of the arrangement of the sands in each of the localities shown on Diagram No. 1, will better illustrate the stratification of the Petroleum Measures than any written description of the same. The vertical range of these sections will be restricted to the total thickness of the Oil-Sand Group or Petroleum Measures to avoid confusion with the overlying Mountain-Sand Group, or Barren Oil-Measures.

Our personal experience in endeavoring to get a correct idea of the stratification of the oil rocks from published well-records, given promiscuously by different drillers or well owners, and colored by their individual theories or pecuniary interests, discourages us from introducing that kind of record here. To the ordinary reader, when given in figures, and complicated with the varying thicknesses of the sands, and the different well-depths caused by undulations of the country or surface elevations of well-mouths above tide, such records become simply unintelligible. To the geologist, depending on the driller's judgment of shale, slate, " soap-stone," " limestone," " flint," " granite," &c., they are a delusion and a snare. It is better, therefore, to present a few *typical sections* in each locality, made from some reliable record which best expresses the stratification there. Twenty such sections are grouped on plates 4, 5, 6, pages 25, 27, 29, and will be described without further reference.

Vertical Sections Nos. 1, 2, 3, at Rouseville, Tarr Farm, and Petroleum Centre.

It will be observed that the Stray is not noted in Nos. 1 and 3; while about midway between them, at No. 2, it appears in more than average thickness. This may be owing to its imperfect development at Nos. 1 and 3, or it may be an omission of the drillers. The strata between sands is seldom examined carefully; and as the Stray of these localities was of no value in the record of a well, when it was a fine gray-sand and the three sands occupied their regular places in the series, its absence in any given record need cause no surprise.

The thinning out of the Second sand at Petroleum Centre is also worthy of notice as a confirmation of the theory that the sands lie in lens-shaped masses and not in continuous belts or broad unbroken bands of uniform thickness.

Vertical Section No. 4, at Shamburg.

Three sands, regular; no Stray. Whole section similar to No. 3; but the Third Sand much thicker; coarse and pebbly; and producing largely a light green oil.

Vertical Section No. 5, at East Shamburg.

Three sands and Stray. Same type as No. 2; but both the Stray and the Third are productive; black oil from the Stray; green oil from the Third.

Vertical Section No. 6, at Pleasantville.

North-western edge of the "Black oil belt." Same type as Oil Creek, three sands and Stray. A few rods to the north-west no Stray is to be found. Here it is a thin, fine grained, gray sand, and unproductive. The Third, although of good thickness, is also fine and compact, and while showing abundant evidences of green oil, does not produce in paying quantities. Gas quite active.

Vertical Section No. 8, at Pleasantville.

Centre of " Black oil district." Three sands and Stray. Compares well with No. 6, except as to Third Sand. Upper part of Stray has coarse yellow pebbles; Production large; Oil dark— ("black.") Third Sand thin, very fine, micaceous and muddy; trace of green oil, and often heavy flow of gas.

Vertical Section No. 7, at Colorado, Hill Farm.

Fair type of a large number of wells on that lead. General stratification about the same as at Rouseville.; but less space between sands. Third sand a beautiful white pebble rock, coarse, porous and well stored with green oil.

Vertical Section No. 14, at Irvin Farm.

About two miles north-east of No. 7. The first record found on this line showing a divided Second sand. First sand, if correctly given, thin. Third heavy; large white pebbles, loosely cemented together with fine sand, but porous, and fairly productive; oil green.

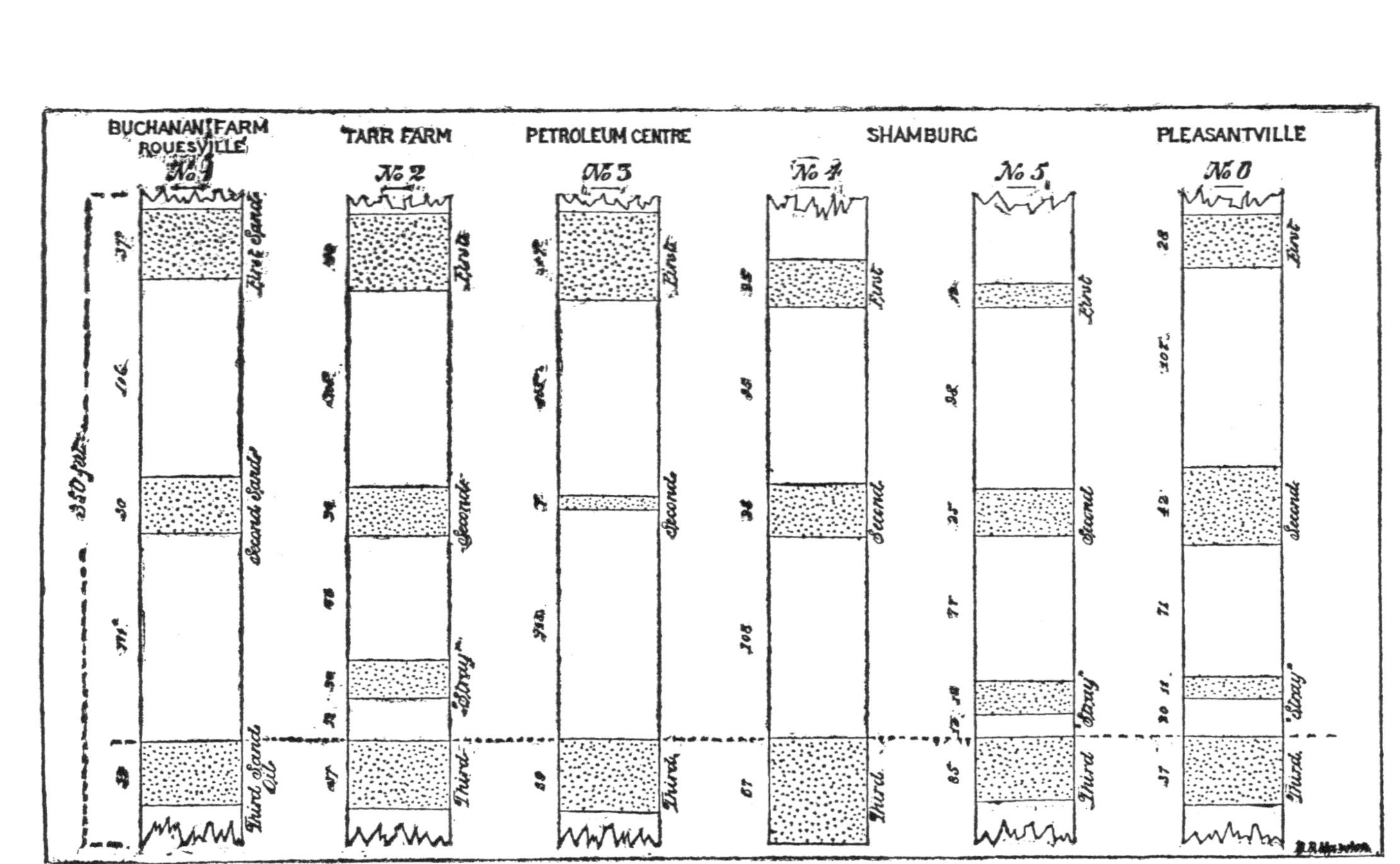

BUCHANAN FARM
ROUESVILLE
No 1
TARR FARM
No 2
PETROLEUM CENTRE
No 3
SHAMBURG
No 4
No 5
PLEASANTVILLE
No 0
350 feet
First Sand
Second Sand
Third Sand
Oil
First
Second
Third
"Stray"
First
Second
Third
First
Second
Third
First
Second
Third
"Stray"
First
Second
Third
"Stray"

Vertical Section No. 9, at Bean Farm.

Between Pleasantville and Cashup. Divided Second. Here the Stray is the oil-producing rock. Wells small but lasting. Oil not quite as dark as at Pleasantville. Third Sand fine, gray, muddy; no oil, and but little gas.

Vertical Section No. 10, at Cashup.

Same type as No. 9. Stray very fully developed, a coarse yellowish sand. From this the oil is produced; in appearance the same as the Pithole and Fagundus oils; not so light a green as the Oil Creek oil, nor so dark as the Pleasantville oil. The wells were large, but short lived, the area of the deposit being limited. The Third Sand, found below, consists of thin layers of fine-grained, gray clayey sands, worthless for either oil or gas.

Vertical Section No. 11, at Pithole.

United States well. We have several records of this well and they all agree, having probably come from the same source. But the similarity between this and No. 1 leads us to infer that it was drilled by men who had been employed on Oil Creek; and not understanding at that time about the Split Second sand, they considered one part of it as a " shell" only, (the name applied to any hard stratum in a well,) and that they arranged their record to suit their ideas of the proper positions of the sands. The well, in being drilled, passed through one mountain sand, (not shown in this section No. 11.) Hence the nomenclature of *four sands* at Pithole. A few rods below this well (see next section No. 12,) and also a few rods above it on the hill side, the Second Sand is found divided; and it seems quite probable that it was divided here also, although it may not have been; for this sand is subject to rapid changes all through this section, sometimes the upper member being the heaviest, sometimes the lower. The history of this well is too familiar to require repetition. It produced largely, with many others in that locality. But the sand rock was thin, and the wells comparatively short lived, partly owing to the rock, and partly to an excess of drilling. Oil, dark, and said to resemble a mixture of Shamburg and Pleasantville oils, " half and half."

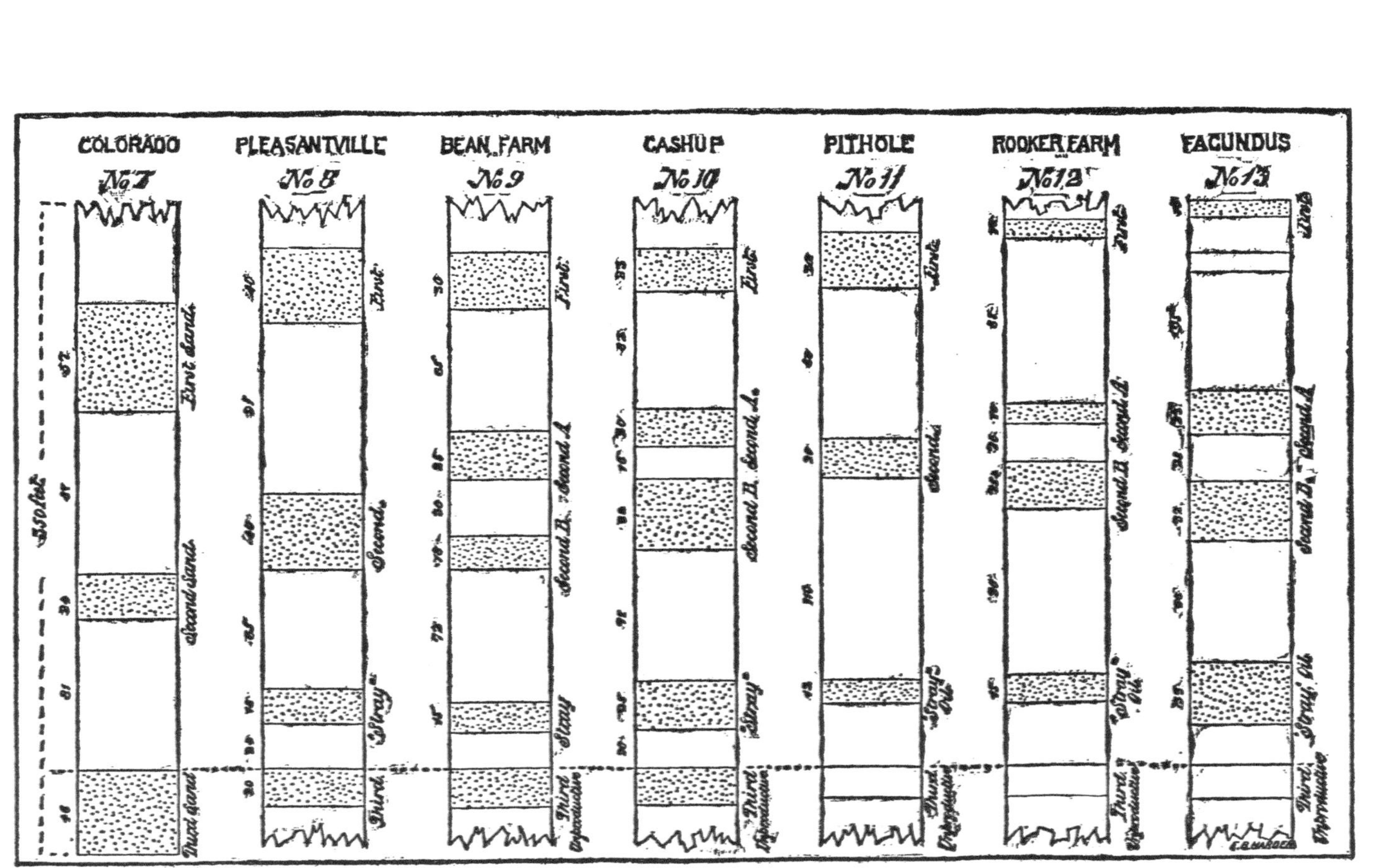

COLORADO No.7
PLEASANTVILLE No.8
BEAN FARM No.9
CASHUP No.10
PITHOLE No.11
ROOKER FARM No.12
FACUNDUS No.13
550 feet
First Sand
Second Sand
Third Sand
First
Second
Third
"Stray"
Second A.
Second B.
Third Unproductive
"Stray" Oil

Vertical Section No. 12, at Pithole.

Rooker Farm. First and Second strikingly dissimilar to No. 11, although but a few rods to the south. These sections illustrate the changeable character of the upper sands of Pithole, to which reference will be made again when the Second Mountain Sand comes under description.

Vertical Section No. 13, at Fagundus.

It was with difficulty that we obtained a complete record of all the sands here, the depth to the oil bearing rock being all that was noted and preserved. The one record secured gives us the first indication of a Split First Sand. The remainder of the section is strikingly similar to Cashup, (No. 10.) The oil is of the same general character as at Cashup and Pithole, and comes, we believe, from the same sand—the Stray. Of one or two wells sunk to below this sand, we could only learn that they found nothing but a number of "hard shells." This is not at all surprising, when we consider how rapidly the Third Sand fines away on the river below Tidioute.

Vertical Section No. 15, on Dennis Run.

Well on Bluff a short distance from the Allegheny River. First Sand regular. Second divided. Third coarse white pebble. Green oil.

Vertical Section No. 16, on Tidioute Island.

One of the early wells. Record, common report, but no doubt resonably correct. First and upper member of Second sands lost by the erosion of the valley. Lower member of Second said by some to have been found at a depth of 18 feet. Third very thick.

Vertical Section No. 17, at Tidioute.

Richardson farm southerly from No. 16. First Sand probably lost by erosion. Second split as in No. 15. Third very thin. Well small.

These sections illustrate the variability of the sand at Tidioute. Along a certain line running from the centre of the river to Triumph, the sand is the thickest found in the Oil Regions, the thickness said to range from 80 to 120 feet. On either side it thins

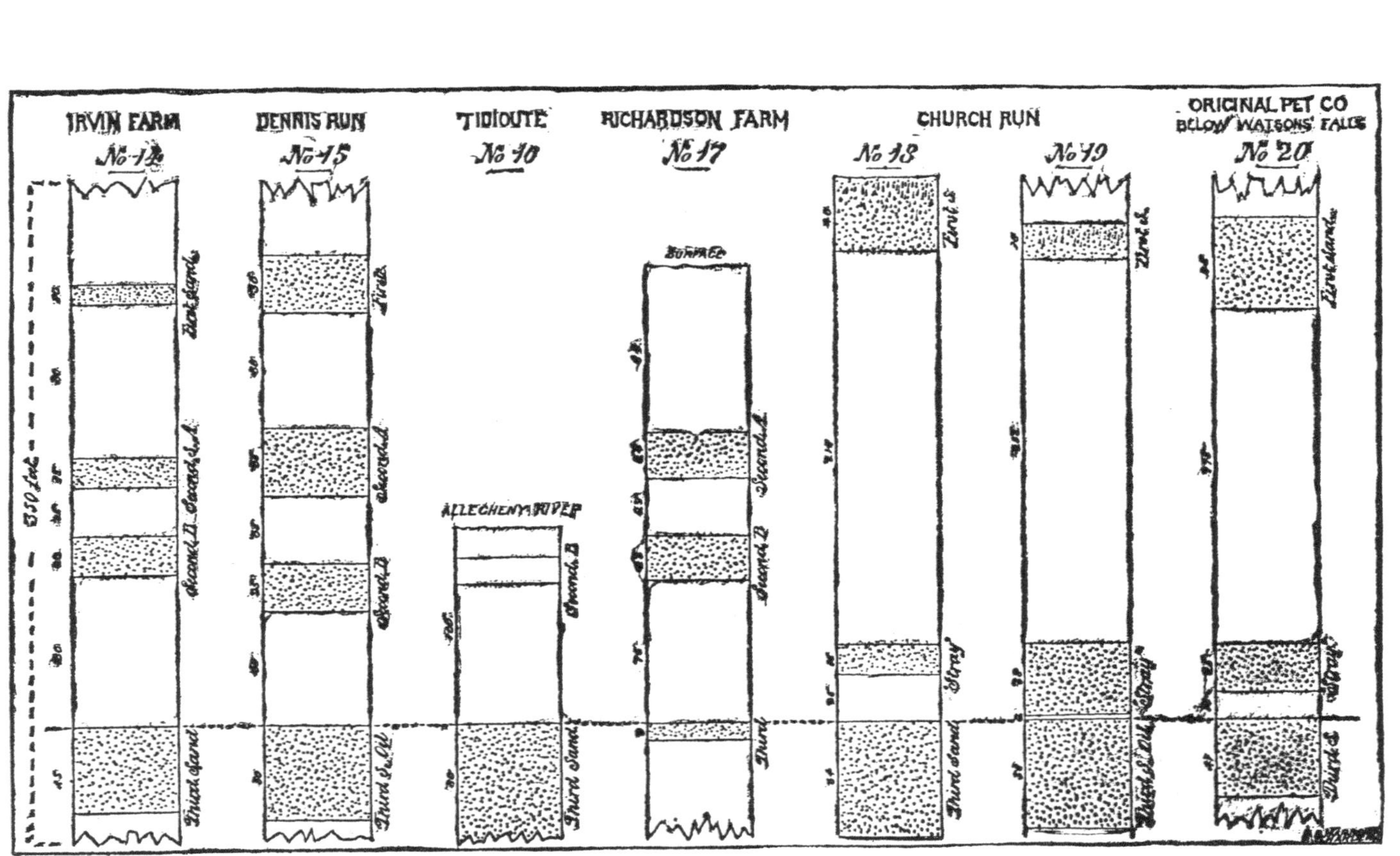
IRVIN FARM
No 14
DENNIS RUN
No 15
TIDIOUTE
No 16
RICHARDSON FARM
No 17
CHURCH RUN
No 18
No 19
ORIGINAL PET CO
BELOW WATSONS FALLS
No 20
350 feet
ALLEGHENY RIVER
First Sand
Second A
Second B
Third Sand
First
Second A
Second B
Third S Oil
Second B
Third Sand
Surface
Second A
Second B
Third
First s
Stray
Third Sand
First s
Third S Oil Stray
First Sand
Stray
Third S

out rapidly, as shown in No. 17, which is probably but little over a quarter of a mile from the centre line.   The great thickness of the oil rock at Tidioute, the small space between it and the Second sand, and the fact of there being a greater dip from this point than from any other shown on diagram No. 1, suggests the possibility of the Stray and Third having joined into one sand.   As No. 5 shows a very near approximation to it, why may not the union have been accomplished here?

Vertical Section No. 18, on Upper Church Run.

M'Guire farm.   An entirely new type.   First Sand thick and lying very high.   Second absent.   Third of more than ordinary thickness; coarse pebble, loosely cemented, and porous.   Wells large.   Oil green.   First instance of over 350 feet from top of First to base of Third.

Vertical Section No. 19, on Lower Church Run.

Parker Farm.   First sand thinner than No. 18.   Second absent. Stray remarkably heavy and almost joining the Third, but non-productive.   Third of good quality and affording green oil in fair quantities.

Vertical Section No. 20, below Watson Flats.

Original Petroleum company's well, a short distance below Watson Flats of Oil Creek, below Titusville.   Same type as at Church Run.   Both Stray and Third to all appearances excellent sands, but unproductive; probably because the territory was exhausted and "watered" before this well was drilled.

This well-record was taken for want of a reliable record on the Watson Flats proper.   The records of Watson Flats, as before remarked, are very imperfect and unreliable.   Nearly all the wells here were drilled previous to 1866, and it is almost impossible to ascertain correctly any of the particulars concerning them.   Consequently we are unprepared at present to attempt to trace these sands, differing so materially from all our other sections, over a broad strip of unproductive territory, to their connection (if they do connect) with the main belt lower down the creek.

Three Types of Stratification.

The sections given above appear to group themselves into three classes.

The first, embracing the central belt on diagram No. 1, shows three sands and occasionally the Stray, and produces green oil from the Third Sand.

The second, covering the south-easterly portion of the field, shows a divided Second Sand, and generally a fully developed Stray, with a fining down of the Third into a worthless sand. This class produces a dark oil over broader and less sharply defined areas than the first.

The third, lying north of the central belt, and embracing Church Run and Watson Flats, shows no Second Sand, but a well defined First Sand, Stray and Third, with green oil from the last.

The causes of these variations in stratification so particularly noticeable in a section (transversely) across the general trend of the oil rocks, will be a most interesting study, when sufficient data have been collected to warrant the undertaking.

Necessity of Thorough Surveys.

It may be thought by some that we are devoting too much time to the particular consideration of so small a portion of the oil field. But paradoxical as it appears, the more the field is narrowed from the region of generalities, the larger and more difficult and important the examination becomes.

It were an easy task to one well versed in the general structural features of the oil-bearing rocks, the direction of their trend, the average dip and characteristic features, to speak in a vague and general way of all the leading facts in relation to them over a wide range of country. But when one confines himself to a small area, and attempts to discover and explain the particnlar agencies that have been employed in the deposition and arrangement of the changing strata of the Petroleum Measures, to line out the possible limits of the oil-bearing strata, and to note the local changes in rock and oil and gas, then the study deepens, and he realizes the fact that safe conclusions can only be drawn from data which he must verify step by step as he proceeds.

In a business so hazardous and uncertain as the production of oil, no *general* information can be relied upon as a basis of continuous success. Many fortunes have been lost by too much confidence in theories crowned by successful operations in one locality, only to be proved worthless by failures at other places.

Nothing but a thorough examination and exhaustive study of *all* the evidences that can be gathered from every source in relation to the oil bearing strata, resulting in specific and detailed information regarding their peculiarities in particular localities, and from which deductions may be drawn enabling the seeker after petroleum to work with more certain results, will be of any practical benefit to him.

Entertaining these views and realizing the liability of error in conclusions drawn, perhaps too hastily, from the partial results of our incomplete surveys, we offer with great diffidence our plan of the dip of the oil-bearing strata and the classification of the sands, explained in this report, with the theories deduced therefrom; and we do not present them as necessarily the precise results likely to be reached in the prosecution of the survey, but rather as present indications of the real direction in which the facts as investigated seem to be leading.

CHAPTER IV.

The Barren Oil-Measures.   Vertical Sections Nos. I. II.

Our upper series of rocks, or Mountain Sand Group, interposed between the Oil Sand Group and the Conglomerate No. XII, has a thickness of about 650 feet as before stated, and is finely exposed for surface examination between Pleasantville and Tidioute.

The valleys of Pine Creek and Oil Creek, and the Allegheny River from Tidioute to Oil City, also cut it nearly to its base, affording a broad opportunity for its geological study as the work of the survey is carried forward.

Our investigations, although confined principally and necessarily to the lines of levels, have afforded, even within these restricted limits, a tolerably good insight into the general structure of the series, the rocks of which are represented on plate 7, page 33, by two additional vertical sections; the one showing the place in the series occupied by the principal persistent sand formations; the other showing the horizons at which fossils have been found.

Vertical Section No. I, is prepared from records of wells drilled on Ennis Hill, Pleasantville.

Vertical Section No. II, has been arranged by referring (up or down the dip) to a central point, the surface levels of all the localities where were found somewhat over 500 specimens of fossil plants, shells, fish and rocks, collected during the season and forwarded to Harrisburg for study and classification in the State Museum.   Although obtained over a wide range of country—from Rouseville to Tidioute, and from Tionesta to Church Run—and at elevations above tide varying from 1,100 to 1,713 feet, they are supposed to be brought from all directions *on the slope of the strata* (as calculated on diagram No. 1, page 17,) to a common centre at Pleasantville.

In attempting a centralization and vertical arrangement of our materials for study in this way, elevating some and depressing others, according to the dip of the country rock and the

3—I.

direction and distance of each particular specimen from the central point—the strictest mathematical accuracy cannot be expected, dealing, as we are, with a series of rocks known to be locally variable. But the effort is serviceable in several respects: for checking, as well as illustrating, the correctness of the observed and calculated dips; for defining more completely the number and order of the sub-formations; and above all for determining the place or horizon of fossil genera and species in the vertical column.

The most careless examiner of these sections will be at once struck with the remarkable similarity of the two sections. A consideration of the widely different methods by which the data from which they are constructed were obtained, will increase his confidence in their truthfulness as representatives of the geology of the Barren Oil-Measures.

A very suggestive feature in the vertical section No. II, is the extra quantity of conglomerate rocks and the abundance of carboniferous fossils which appear in the *upper* part of the series, and their absence from the *inferior* portion of it. This might be attributed to some lack of thoroughness in the search for them along the lower outcrops, were we not re-assured by the evidence of a different class of fossils beginning to appear as soon as we pass, descending the column, the last of the conglomerates, the Second Mountain Sand.

Conglomerate No. XII.

No part of Conglomerate No. XII (the base of the Productive Coal Measures) remains *in situ* on Ennis Hill; but the remnants of it, fragments of sandstone containing impressions of Lepidodendra, Sigillaria, Calamites, &c., are strewn over the hill top; and the character of the " bed rock" exposed in excavations for " conductors" of oil wells, would indicate a close proximity to its former base. This view of the case is strengthened by the fact that the sand-rocks from the hill at Cashup and the River Hills toward Tionesta—which have always been considered a portion of No. XII—although lying at a lower tide level than Ennis Hill, do, when carried into position there on the plan adopted, overshoot the hill-top, and form *in the air* a portion of the Conglomerate as shown in No. II.

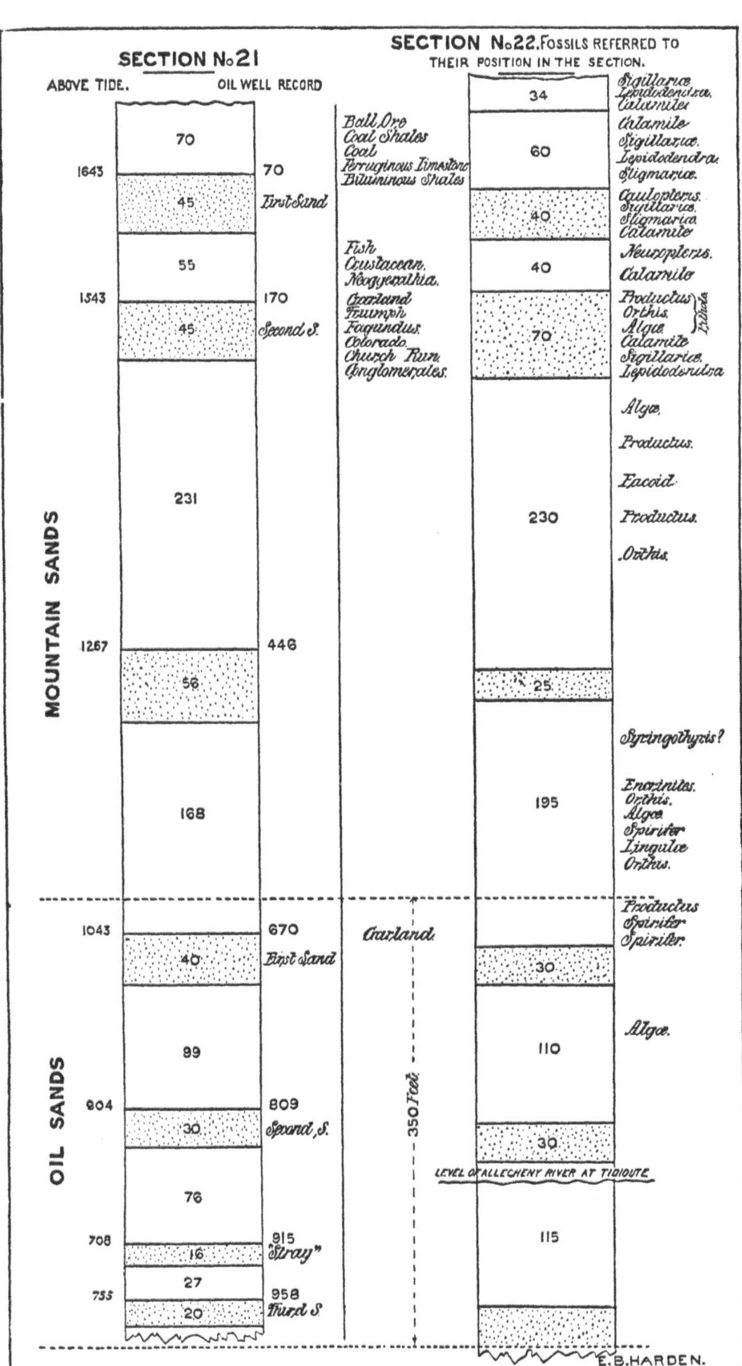

SECTION No. 21
ABOVE TIDE.
OIL WELL RECORD
SECTION No. 22. FOSSILS REFERRED TO THEIR POSITION IN THE SECTION.
MOUNTAIN SANDS
OIL SANDS
70
1643
70
45
First Sand
55
1543
170
45
Second S.
231
1257
446
56
168
1043
670
40
First Sand
804
99
809
30
Second S.
76
708
915
16
"Stray"
755
27
958
20
Third S
Ball Ore
Coal Shales
Coal
Ferruginous Limestone
Bituminous Shales
Fish
Crustacean.
Neogyexalhia.
Garland
Triumph
Faqundus.
Colorado.
Church Run.
Conglomerates.
Garland.
350 Feet.
34
Sigillariae.
Lepidodendra.
Calamites
60
Calamite
Sigillariae.
Lepidodendra.
Stigmaria.
40
Cauloplenis.
Sigillariae.
Stigmaria.
Calamite
40
Neuropteris.
Calamite
70
Productus
Orthis.
Algae.
Calamite
Sigillariae.
Lepidodendra
Lingula
Algae.
Productus.
Fucoid.
230
Productus.
Orthis.
25
Syringothyris?
195
Encrinites.
Orthis.
Algae.
Spirifer
Lingulae
Orthis.
Productus
Spirifer
Spirifer.
30
Algae.
110
30
LEVEL OF ALLEGHENY RIVER AT TIDIOUTE
115
E. B. HARDEN.

In reference to the term "Conductor," or "Conductor-hole," used above, it is well to explain for the benefit of such as live beyond the limits of the Oil Region, that in preparing to sink an oil well, in localities where the original rock formation lies undisturbed at a depth of not more than 10 to 15 feet from the surface, it is customary to sink a shaft to the solid rock, similar to a common water well. This is a "Conductor-hole." The "Conductor," or square box of heavy plank, is then inserted from the derrick floor to the rock, and the drilling is commenced. The lower end of the conductor is "shouldered" in the rock and carefully secured, to prevent the surface accumulations from washing into the well. In doing this it is necessary to dig until a solid stratum is found, and sometimes several layers from the top of the rock will be thrown out before this object is attained. From these excavations we have secured many fine specimens of fossils which otherwise would have been overlooked.

When the rock cannot be easily reached by digging, an iron tube is driven for a "Conductor." On some of our creek bottoms, from 75 to 200 feet of tubing has been required; evidence that these valleys have been excavated to that depth at some former period, and then, by changes in the water levels, have been filled up again with *debris* from the neighboring hills.

Cuyahoga Shale No. XI.

Assuming the Conglomerate No. XI to be correctly located in the vertical section, we find between it and the First Mountain Sand, of oil wells, an interval of about 60 feet. Into this interval will fall, by the method of adjustment adopted, all our specimens of Iron, Coal, Coal Shales, Bituminous Shales and Fire Clays, (characterized by certain fossil plants, especially Lepidodendra and Calamites,) from this district.

Subconglomerate Coal Bed.

The layers (if more than one) of Coal found in this group, are thin, and, as far as known, of no great extent or economic value. Two openings only have been visited—the M'Kissock Bank, now belonging to Mr. S. Q. Brown, on the ridge, south of Cashup and east of Pithole; and another on the Eagle's Farm, about three miles north-east of Plummer. Both of these are now in disuse. The drifts have fallen in, preventing inside examinations, and nobody could be found to give any information concerning them, beyond the fact that the coal was too thin to be profitably worked. Coal is known to exist in several other places north of the Allegheny River, and east of Oil Creek, and is referrable, apparently, to the same place in the section. It is probable that but one small coal bed exists beneath No. XII, and that it represents the subconglomerate coal bed, one foot thick, seen outcropping around the mountains of Bradford and Tioga counties, and along the brow of the Allegheny Mountain.

Upper Berea Grit No. X. First Mountain Sand.

The First Mountain Sand plainly outcrops, in many places not far from our centre (Pleasantville,) in such positions and at such elevations as prove beyond doubt that it must be the same as that shown in the well. (Ver. Sec. I.) The fossils found in it (Stigmaria, Sigillaria, Calamites, Caulopteris, &c.,) when referred by dip to Pleasantville, group themselves within a vertical space of forty feet.

Catskill? Shales No. IX?

Between the First and the Second (or next lower) Mountain Sand, is a reported interval of 55 feet of Shales, (Vert. Sec. No. I,) but appearing somewhat thinner in No. II. In these carboniferous shales are well preserved fossil ferns (Neuropteris,) Asterophyllites, Noeggerathia, Calamites, Stigmaria, &c., with occasionally crustacean and fish remains.

Lower Berea Grit? Second Mountain Sand.

To the Second Mountain Sand is given, by well-records within a circle of comparatively small radius around Ennis Hill, a thickness varying from 20 to 90 feet; which leads us to infer, that it is much altered in its character from where we see it on its distant north and north-east outcrops. It often happens that where a sand approximates in its composition to shale, it is not readily distinguished from shale by the drillers; so that the alleged thickness of the sand depends very much in such cases on their judgment.

The outcrops of this sand we believe to be (after a careful calculation of levels and slopes) the coarse conglomerates which appear in the hill-tops at Garland, Triumph, Fagundus, Colorado and Church Run, as noted in Vertical Section II.*

The fossils are similar to those of the First Mountain sand, of a true carboniferous type, and not found in any rock below this Second Mountain Sand: Calamites, Sigillaria, Lepidodendra. At Pithole this rock contains Productus, Orthis, and Algæ.

Proofs of the Identity of the northern Conglomerates with the Second Mountain Sand.

As this view of the case is contrary to the generally received opinion concerning these rocks, the popular belief being that the abovementioned conglomerates are outliers of the Great Conglomerate No. XII, it is proper here to show a part of the process by which an opposite conclusion has been reached, and for that purpose we append the following figures:

	A.	B.	C
A. Elevation above tide at Garland, &c.,			
B. Correction for dip, referred to Pleasantville,			
C. Elevation above tide at Pleasantville,			
At Pleasantville, Conglomerate No. XII, (base,)	......	......	1,713
At Pleasantville, Mountain Sand No. 2...... "	......	......	1,498
At Triumph, conglomerate.................. "	1,711	175	1,536
At Fagundus, " .... "	1,632	110	1,522
At Garland, " .... "	1,833	304	1,529
At Colorado, " (Ware farm,) ... "	1,623	108	1,515
At Church Run, " (Parker farm,).. "	1,553	40	1,513

Difference in the relative horizons of the two rocks............. 200

The corrections for dip will be understood by reference to diagram No. 1, page 17.

While we cannot claim for the above calculation an accuracy admitting of exact demonstration, the dip being necessarily more or less subject to local changes on account of the slight undulations before spoken of, we submit that the fact of these rocks agreeing so nearly in tide-datum-level when traced to Pleasantville on the same slopes which have been seen to belong

[* If Mr. Sherwood be correct in tracing the Olean Conglomerate (Garland) through New York to Tioga county, and there identifying it with the top layers of the Chemung (VIII,) we must accept the conclusion that Coal-Measure life-forms began to appear in Middle Devonian times; for the Catskill Red mass is undoubtedly our equivalent for the Old Red Sandstone of England. J. P. L.]

to the Oil-Sand Measures, should be considered satisfactory evidence of their inclusion in one common geological horizon. Nothing but a marked unconformability of strata could throw either one of them into Conglomerate No. XII, which lies about 200 feet higher.

Further Proofs, from Oil Wells.

That these conglomerates are not the same as No. XII may also be proven by comparing our levels with the records of wells drilled at some of the localities named.

If the Carboniferous Rocks lie conformably on the Petroleum Measures—and there has never been assigned a reason to doubt that fact—then, speaking in a general way, the plane of Conglomerate No. XII must be *approximately* parallel to the plane of the Third Oil Sand, and all wells sunk from the base of No. XII to the top of the Third Oil Sand should be of nearly equal depth.

Let us see the result of this test:

```
                                                                     Feet.
Depth of well at Pleasantville; from base of No. XII to top of Third Sand, 958
At Triumph; from base of Triumph conglomerate to      "        "      755
At Colorado;        "       Colorado      "           "        "      767
At Church Run;      "       Church Run    "           "        "      764
At Fagundus;        "       Fagundus      "           "        "      ...
      "             "           "         "        to top of Oil-produ-
                                                      cing Sand.......... 701
```

These figures are all from records of wells located near the base of the outcrops in each place, so that they may be considered as nearly as possible absolute measurements from the base of the one rock down to the base of the other.

Here then we have:

```
                                                                     Feet.
From No. XII to Third Oil-Sand in Pleasantville........................ 958
From Triumph, Church Run and Colorado conglomerates to Third Sand,
  an average of...................................................... 762
                                                                     ─────
  Difference in the relative horizons of the two rocks.............. 196
```

As this result agrees so nearly with that obtained by the first method it must be accepted as good corroborative evidence that these conglomerates are not outliers or outcrops of the Great (Seral) Conglomerate No. XII. And we think it is also an argument in favor of the conformability of the strata of the two systems; since, if they were unconformable, the change of levels could hardly be supposed to occur in such a manner as to allow the wells at the places named to show records agreeing so remarkably in their measurements from the Conglomerate down to the Third Sand.

Inference in relation to the Fagundus Oil-Sand.

The depth of wells below conglomerates also affords incidentally another proof in support of the opinion that the Fagundus oil rock is the "Stray," and not the "Regular Third" of Oil Creek.

For, if we are correct in classing the Fagundus conglomerate with those of Triumph and Colorado, then a well at Fagundus should be as deep as one at either of the other places, if they *all* obtain oil from the Third Sand. But we find it less deep by *over* 60 *feet*, which is about the distance discovered to be wanting (see diagram No. 2, page 19) to complete the symmetry of the general dip, as shown on diagram No. 1; and it can only be accounted for on the theory that the Oil rock of Fagundus is the Stray, and lies above the plane of the Regular Third.

Characteristic variations of the Second Mountain Sand.

In tracing the Second Mountain Sand, or its equivalents, from its best developments in the places above named, we find, as we proceed south-easterly, a gradual fining down of its particles, and a change from coarse and massive gravel to thin bedded argillaceous sand. At Pleasantville it has already taken on something of this form; while, at Pithole, in the bluff wells, where it should appear, it is not mentioned at all; and it seems probable that it is here divided into two thin bedded, laminated sands; one lying above and the other below the plane corresponding with its position in the section further north-west. The valley of Pithole Creek is eroded to a considerable depth below the horizon of its base; and yet the sand does not show itself with any prominence along the bluffs. In addition to these evidences of having lost its north-western character, we have collected a number of specimens dug from " conductor holes," at about the level where this rock should be found. These specimens consist of thin plates of fine-grained sand, showing wave-marks and well preserved fossil shells, (Productus, Discina,

Orthis,) as well as sea-weed (Algæ) and other evidences of a deposit in quiet water, at a distance from the currents which must have swept along the pebbly accumulations of the north-west.

Bedford and Cleveland Shales No. VIII.

Beneath the Second Mountain Sand or conglomerate, the base of which may be considered as lying about 200 feet below the base of the Great Conglomerate, No. XII, all the fossils found are Subcarboniferous, consisting entirely of marine shells, sea-weeds, &c. Thus far not a single specimen of the Carboniferous land-vegetation so abundant in the formations last described, has been secured.

The names and positions of some of the fossils collected in this part of the series are shown in section No. II.

No attempt has been made to number the Mountain Sands below the Second, as it would be of no practical use. Except on a narrow range overlying the " green oil belt," they lie with great irregularity, especially in the south-east part of our district. The well-records in that direction exhibit them changing level, splitting up and fining away ; until in some places they so nearly approximate the character of shale, and become so interstratified with shale, that it is almost impossible for the drillers to distinguish one from the other.

This explains why some of the wells record such a number of sands. As many as seven are noted in the space of 170 feet in one well on Stewart's Run.

CHAPTER V.

Unreliability of Well Records.

Well informed men, unacquainted with the method of drilling, have supposed that the stratification of the rocks could be accurately estimated from a study of the productions of the sand-pump. When there is a sudden and marked change from sand to slate, or from slate to red-shale, and the measurements are accurately made, this may be true. But in many wells the bands of sand are so fine-grained and thin, with shales and slates of nearly the same color so frequently intercalated, that the drill passing from one to another, and grinding all the particles together before they are brought up in the sand-pump, renders a correct report of their original stratification an impossibility. To this may be added other causes of unreliability:

First. Careless sand-pumping; in which case the loose sand, remaining in the bottom of the hole, is re-drilled, follows the tool down, and may show itself for several feet after the sand-rock has been passed.

Second. Falling of particles from above, while running the tools and sand-pump up and down; which often happens in sufficient quantities to change the character of the drillings.

Third. Careless measurements of these intermediate changes in the rock; and indeed they are seldom measured at all, being estimated by the *number of "screws" run since the last measurement in the sand;* that is, the number of screws run in shale after getting through a sand.

These remarks apply, however, more particularly to the old wells, before drillers had acquired that nice discrimination which enables them now to judge with considerable accuracy of the quality of the rock being drilled by the manner in which the tools make progress upon it; and before the steel wire, now made expressly for measurements, came into use.

Screw Measurement.

The " Temper screw" is an instrument by which the cable attached to the drilling tools is connected to the " walking beam."

It is unscrewed gradually as the drill penetrates the rock. If there was no " stretch" in the cable, the depth drilled could be told with considerable accuracy by the length of screw let out. But with a thousand feet of cable in the well it may, under some circumstances, stretch faster than the drill penetrates the rock, and the screw has to be " taken *up*" while the drill is actually proceeding *downward*. From this the uncertainty of this kind of measurements will be easily understood.

In drilling, a " centre-bit" and " reamer" are used. The centre-bit is a steel-pointed bar of iron, say three inches in diameter and three feet long, with a " pin" at the upper end by which it is screwed to the " Auger stem." The point is flattened and drawn out wedge shape like an ordinary stone-drill. This drills a hole say 4½ inches in diameter—and in ordinary drilling can be run the length of the temper-screw, or about 3 feet, without being withdrawn to be " re-dressed" or sharpened.

When this is drawn out there remain in the bottom of the well the finely-ground particles of rock from a hole 4½ inches in diameter and 3 feet deep. This is removed by the sand-pump as a thick mud; and if poured into a pail and washed for the sediment—allowing the depth drilled to have been partly of sand and partly of softer material—little except sand will remain. If not washed, the sediment dries into a mud cake defying any correct judgment of its original character.

The " reamer" is now introduced to enlarge and smooth the bore. It is a little shorter than the bit, and instead of being sharp is about 2½ inches broad on the bottom, and " dressed" to the curve of a circle of 5½ inches diameter. Its work is to cut down the " shoulder" of ½ an inch, left by the centre-bit, between the 4½ inch hole drilled and the 5½ inch hole required. Of the 3 feet bored by the centre-bit we will suppose the first foot to have been sand and the remaining two feet slate or shale. The " reamer" commences its work, and as it pounds away on this ½ inch ring, the " spawls" of sand are knocked off and fall to the bottom. By the time the sand is cut through, and a few inches are gained in the shale, the chippings from above have filled the bottom of the hole, and the reamer, taking a bearing on the sediment, begins to tamp it down. When the reamer ceases to make progress it is withdrawn and the sand-pump is again introduced.   As this can remove only the loose sediment, the bottom of the centre-bit hole remains tamped full of sand.   This is re-drilled by the next bit (which perhaps reaches another stratum of sand before being withdrawn) the sand-pumpings are washed, and the sand-rock reported and recorded.'

In this way, no doubt, in localities where the sands and shales alternate frequently, many of the well-records are *unintentionally* falsified, and a very different stratification is reported from that which actually exists.

Strata between Oil-Sands.

The above digression on well records explains one of the reasons why we have made so little definite mention of the strata between sands.   Our surface examinations have not been sufficiently extended and thorough to warrant it, and the well records afford no trustworthy information respecting them without an almost impossible amount of labor in sifting out the reliable from the unreliable.   It seems more prudent therefore, for the present, to pass them with only a general notice.

The Red Beds.

One feature, however, is too important not to be carefully mentioned.   It is worthy of special attention.   Bands of red shale (" Chocolate" or " Putty Rocks") occur between the Second and Third Oil Sands of Oil Creek.   They are reported absent from the wells at Tidioute; but red is the predominating color of a major part of the Tionesta well drillings.   Below the Third Oil Sand we believe similar red rocks appear universally.

What their real significance is in relation to the oil sands is not yet fully understood; but there is no doubt that a better acquaintance with them will be of great service to the oil operator, as one of the indications distinguishing productive from unproductive territory; or perhaps, in some places, as an indication that the horizon of the Third Oil Sand has been passed.

[It seems hardly possible that these red beds represent the red Bedford Shale of Ohio, immediately underlying the Berea Grit. See Dr. Newberry's Reports, discussed in Article III, of this Volume.   Did they lie between the Second and Third *Mountain* Sands no one would hesitate to call them the Bedford Shale.

But they lie several hundred feet lower, between the Second and Third *Oil* Sands, and below the Third Oil Sand. They will perhaps hereafter be identified with the Mansfield red beds (of the Survey of 1841) crossing the Tioga River. The red beds of the Clarion wells below Ridgeway may be the same; but they are more likely to be proved by our future work to be the Catskill No. IX of the front face of the Allegheny mountain, lying far higher in the series than the Mansfield red beds of VIII.—J. P. L.]

CHAPTER VI.

Garland Conglomerate, and Hosmer Run Oil Rock.

A passing allusion has been made to the conglomerate capping the hills at Garland. The elevations at this point were taken in connection with the old oil-pits and wells on Hosmer Run, mainly with a view to ascertain the position of that oil rock in our series. The levels run from the nearest railroad station, and not connected with our own base lines, require some future adjustment, but are accurate enough for the present purpose.

These Oil Pits, twenty-five or more of which are distinctly to be seen along the bottom of the run, were dug, like those on Oil and French Creeks, long before the settlement of the country by the white man, and there remains no trustworthy tradition of their origin.

During the early excitement produced by the discovery of petroleum a number of wells were put down in the neighborhood of these ancient pits, but none of the wells proved remunerative, although almost all of those near the pits yielded small quantities of oil. More oil might be dipped from the " conductor hole," in preparing to sink a well, than could afterwards be pumped from the well when completed.

One well, put down to a depth of 1,060 feet, found no clearly defined sand after passing the first, which is said to lie about fifty feet from the surface of the bottom-lands of the run. This first sand is reported to be a good pebble rock, containing heavy-gravity oil.

Many other wells were put down on this Hosmer Run, on the Brokenstraw, on the Little Brokenstraw, and on Spring Creek,— to depths varying from 300 to 900 feet,—all of them proving, as far as can be learned, the existence of but one oil-bearing sand in that district, and that not sufficiently productive, where developed, to warrant the expenditure of capital in sinking oil-wells.

Based on levels already referred to, we have ventured the opinion that the Garland conglomerate capping the hills around the Hosmer Run wells, belongs to the same plane as the conglomerates of Colorado, Church Run, &c., and corresponds with the Second mountain-sand of Pleasantville.

If now it can be also shown that the Hosmer Run oil rock corresponds with the First Oil Sand of our Petroleum measures, and that the Hosmer oil rock and the Garland Conglomerate lie relatively to each other as do the First Oil Sand and Second Mountain Sand to each other, then the probable correctness of our classification will be confirmed or greatly increased.

1st. *As to the Hosmer Oil Sand.*—Allowing the Hosmer Run oil rock to have the same dip that we have shown the Petroleum measures to have, we should look for its representative in Pleasantville at a depth of about 630 feet. This is about 40 feet too high to agree with the First Oil Sand shown in vertical section No. 1, and would place its *base*, (calling the Hosmer Sand 40 feet thick,) on a level with the *top* of the First Regular Oil Sand.

2d. *As to distance between conglomerates and sands.*—Supposing the Garland Conglomerate to be Second Mountain Sand and the Hosmer Run oil rock to be First Oil Sand, then the distance between them, by the vertical section No. 1, is about 455 feet, and by our levels at Garland 410 feet.

This want of exact agreement, considering the distance between the points of observation and the liability of local undulations in the general dip, can hardly be considered a cogent argument against the proposed identifications.

The only conclusions, then, that can be drawn from the above data are these:

1st. The Garland conglomerate is the equivalent of Mountain Sand No. 2 of the Barren Oil-Measures.

2d. The Hosmer Run oil rock corresponds with the Oil Sand No. 1, of the Petroleum Measures.

3d. The Second and Third Oil Sands are wanting in that more northern locality.

CHAPTER VII.

Drift.

In addition to the local drift composing the " flats" of nearly all the streams in this district, there are in many places large deposits of Drift of an entirely different character. The latter are more noticeable in localities within the range of the larger streams flowing from the north, and they contain many specimens of rocks which could only have come from a remote distance. Some of these accumulations, of considerable depth, may be seen on ridges elevated from 150 to 200 feet above the present streams.

On Magee Hill between Pleasantville and Titusville, a well sunk near what was supposed to be the base of one of these Drift deposits, and within 50 rods of the plain out-croppings of undisturbed strata, required 70 feet of pipe to reach the bottom of the Drift.

No large amount of this Drift has been noticed south of the Pleasantville ridge; but small scattering boulders of foreign rock may be found even on the slopes of Ennis Hill, the highest point in Venango county.

These facts suggest another interesting field of study in this district.

CHAPTER VIII.

Clarion and Butler Counties.

A desire to confine this report to such matters only as have come under instrumental observation during the past season, and to such facts as have been examined with considerable care— prevents any attempt to connect the work in Venango with the oil fields of Clarion and Butler. We have not had the opportunity of even visiting these districts since the survey commenced. It would be unwise, therefore, to endeavor at present to trace our oil sands further in that direction. As well might a traveler in an unexplored country, as he sketches at evening the scenery and incidents of the day's journey past, add in advance an account of to-morrow's explorations. He may have been traveling in a country so comparatively uniform and unvaried in its general physical features that he would be warranted in the expectation of a continuance of the same characteristics as he proceeds. But an intervening mountain, a sudden curve in a river, may hide from his view a landscape entirely new and unexpected.

It may be said, however, without much probability of error, that the examinations of the "Upper Oil Belt," taken in connection with the general dip of the strata and numerous facts incidentally ascertained in relation to Clarion and Butler, lead to the belief that the oil-bearing sands of those districts, when surveyed and classified, will be found to belong to the same geological horizon as those of Venango.

CHAPTER IX.

M'Kean County.

For the same reasons, also, nothing definite can be said of the oil wells in M'Kean county. Here the relation of the oil rocks to those in Venango remains somewhat doubtful. The depth at which the oil is obtained would appear to indicate a lower band of oil-bearing sand than any found in this district. But in the absence of reliable data to work from, it would be futile at present to venture an opinion on the subject.

The very recent oil flow from Mr. Beattie's Gas Well at Warren, comes from a horizon at least 600 feet below the lowest (Third) Oil Sand, and adds unexpected force to the opinion hazarded above.

CONCLUSION.

In the foregoing pages we have aimed simply to give a report of the progress of the work of the Second Geological Survey of

Pennsylvania in the Venango district, accompanied only by such historical and descriptive facts in relation to oil-well drilling as seemed necessary for explaining the subject

The daily exactions of field-work have not only precluded the possibility of extending our observations beyond the narrow limits of our own special investigation, but deprived us also of many facts in relation to the district surveyed, which could only have been obtained by an expenditure of time in other directions not to be afforded, under the circumstances.

Many inviting avenues of investigation have opened in this comparatively new field of geological study ; and it would have been interesting to follow some of them to practical conclusions. But in view of the continuance of the survey, it is well perhaps that no partial examinations should be made in advance of the instrumental work, theories based on which might not be sustained hereafter.

If a feeling of disappointment is felt at the incompleteness of this report, it should be remembered that it is intended only as a brief account of the progress of the work during the first few months of its prosecution, and that no conclusions of very visible practical importance ought reasonably to be expected until the work has gone much further forward.

JOHN F. CARLL.

PLEASANTVILLE, PA., *Feb.* 1, 1875.

II.

OBSERVATIONS

OF THE

GEOLOGY AROUND WARREN, PA.

BY F. A. RANDALL.

WARREN, *March* 3, 1875.

To Professor J. P. LESLEY,
            State Geologist:

SIR:—After a longer delay than I intended, I forward to you my notes, and you can make what disposition of them you think best.   I feel somewhat distrustful of my elevations, owing to the imperfection of my instruments.   As soon as the weather permits I shall go over the ground again.   *   *   *

Very respectfully yours,

F. A. RANDALL.

The District examined.

The region covered by these observations is on the north-western edge of the Sixth Bituminous Coal Basin of Western Pennsylvania, and extends from a point four miles north of Warren,—thence south-eastward to Great Bend, two miles west of Kenzua Creek mouth,—thence west along the P. and E. railroad twenty-four miles to Garland.

The average dip to the south-west (with frequent minor undulations) I make to be 20 feet per mile.

Oblique lamination (or false-bedding; or current-bedding) in the coarser sandstones and conglomerates is *always towards the north and north-west.*

Order of formations.

The highest rocks of the region are to be found towards the East.; they are eroded away towards the West.

The highest pile of formations is to be found, therefore, at Great Bend, east of Warren, the hill tops there reaching an altitude of about 2,100 feet above tide level; as shown in the long section attached to this report. Descending from the hill tops the following series of Sandstones, Shales, &c., are seen outcropping horizontally one beneath the other:

a. Sandstone, 50 feet,   } Coal Measures, XII.
b. Conglomerate, 10 feet, }
c. Shales; (Coal,) 40 feet.
d. Sandstone, 120 feet.
e. Conglomerate, 40 feet, (First Mountain Sand.)
f. Slate rock; fine Sands; one conglomerate. 200 feet.
g. Sands and Slates; Shales. (*Chemung*) VIII. 360 feet.

But to the West of Warren, at Garland for example, the conglomerate *e* caps the hills.

The thickness of each member of the above series constantly varies; the conglomerate in *f*, for instance, being 15′ at Garland, 50′ on Hatch's run, 40′ at Great Bend, &c.

Formations described.

a. Coal Measure Sandstone; exceedingly compact and white in its lower layers. It may be considered as graduating downwards into and forming part of *b.*

b. Conglomerate, No. XII,? compact, and resembling the Conglomerate of the Anthracite Coal Basins of Eastern Pennsylvania, much more than it does the conglomerates *e* and *f* below.

c. Shales, containing near the bottom, from one to three thin (Sharon) Coal Beds which occupy local depressions or hollows in the upper surface of *d.* [This Shale formation represents the Cuyahoga Shales of the Ohio Survey; the Umbral of Roger‐ Final Report; and No. XI Red Shale of Middle Pennsylvania.

d. Sandstone, friable, coarse-grained, containing undoubted Coal Measure plants. [It is probably Mr. Carll's First Mountain

Sand, with its underlying shales and the upper Berea Grit of Ohio.]

e. Conglomerate, forming cliffs, broken by clefts and fissures, bear-caves, &c., around the hill. Huge angular blocks of it lie scattered about the higher terraces. Being forty or fifty feet thick, very solid and massive, and composed, especially at its base of large quartz pebbles in a cement matrix of coarse sand, somewhat ferruginous, it makes a great show about the country to the North into the State of New York, where it forms the Rock Cities of Cattaraugus and Chautauque counties and caps the hills near Ellicottville. [It has, therefore, been mistaken for the Great Conglomerate No. XII at the base of the Coal Measures. It is, in fact, Mr. Carll's Second Mountain Sand, 200 feet below No. XII. And it may represent the bottom member of the Berea Grit of Ohio. The two, *d* and *e*, correspond in geological position exactly with the great Sandstone Formation No. X (Vespertine) of Middle and Eastern Pennsylvania, outlying patches of which, 500 to 1,000 feet thick, form the higher knobs of the Catskill Mountain, between the Delaware and Hudson Rivers. The only objection to this identification at present comes from a supposed tracing of the conglomerate *e* through New York to Tioga county, Pennsylvania, to where it seems to *underlie* the Red Catskill. In which case the Red Catskill (IX) or Catskill proper must be identified with the bottom layers of the sandstone *d.*]

f. Micaceous slates; subordinate beds of sandstone, fine grained, buff colored and red; with one conglomerate composed of numerous layers of coarse pebble rock, making perpendicular bluffs and cliffs (lower down on the hill sides than the cliffs of the upper conglomerate *e*) capped by thick beds of exceedingly hard slate. Formation *f* as a whole makes rounded and angular isolated hills with truncated tops, resembling Indian Mounds, or unfinished pyramids, and well defined terraces frequently appear upon their sides.

[The red layers, and the conglomerate layers, both remind us of the red and pebble layers in the Catskill No. IX of Middle and Eastern Pennsylvania. In Mr. Carll's Venango county sections occurs a Mountain-Sand rock about 400 feet below the XI Coal, and this may be the Warren conglomerate in *f.*]

[At 360′ above the river Mr. Randall locates his shale beds full of *Spirifera*, and marks them as Limestone.]

g. The Chemung Oil-sands and shales occupy the hillsides of the Allegheny River Valley to a height of 500 feet; and below the river bed for many hundred feet the Beattie and other wells have passed through similar shales of No. VIII (Chemung Formation.) The hillsides exhibit outcrops of shales, slates and sandstones, with occasionally interstratified impure limestone layers, the lime having been derived from the shells of *Spirifers* with which such layers are crowded. There beds of conglomerate also occur. But none of these yield oil. Yet they are the northern extension of the three or more Oil-Sands of Venango County, carried up above water level by the general rise of all the measures towards the north. The upper beds of the series are chiefly sandstones and micaceous slates; softer shales predominate on the slopes descending to the river bottoms. The first 200 feet of the Beattie well went through nothing but soft shales.

[No discussion of the oil struck at a greater depth in this well would be worth while at present, nor until a careful survey can be carried east and north into M'Kean County. Nor can the identification of conglomerates be more than indicated as above, until far more careful and elaborate measurements are made. More definite results will, no doubt, be reached this year.]

Fossil Life.

The Coal Measure plants of the uppermost 200 feet, viz: *Sigillaria, Lepidodrendron, Calamites*, are very abundant in the sandstone *d;* and no other but these air-breathing trees and reeds are to be seen above the conglomerate *e*. [Here Mr. Carll finds his fish spines, *always in loose pieces of red sandstone* 4″ or 5″ thick, in or at the top of his First Mountain-Sand.]

In the middle 200 feet of subcarboniferous rocks, on the other hand, none of these Coal Measure plants appear. The change is abrupt at *e*. Sea-weeds become now abundant;—Polyps: *lithostrotion;*—Star-Fish crinoids: *Cystidea, Archæocidaris,* ————; *Pentremites,* (rare);—Shell-fish (mollusks): *Productus, Chonetes,*

*Strophomena, Rhynconella, Orthis, Syringothyris,** *Platyceras, Aviculopecten, Orthoceras, Bellerophon;* all Subcarboniferous forms.

The Devonian (Chemung) fossils of the lowest 500 feet, are again for the most part different, although Sea-weeds are very abundant; and, of course, Lepidodendra and Equiseta are rare, being merely floated trunks; Crinoids are also rare; Mollusks are abundant: *Lingula, Discina, Spirifer, Productus, Rhynconella, Goniatites, Euomphalus, Bellerophon,* and many other species of conchifers;—Crustaceans: *Trilobites;*—Fishes: *Holoptychius* and *Coccosteus.*

Section at Warren.

[Mr. Randall has been a diligent observer and collector at Warren for many years, and possesses a cabinet of value, in which are represented numbers of all the generic types of life above mentioned, with some not yet described.

He has kindly prepared a section of the hill country through which the Allegheny River flows at Warren.

This section has been re-drawn to a natural scale, (horizontal and vertical the same,) and is now published with this report; reduced by photo-lithography to a scale of 1,200' to 1".

The series of formations above described are shown by it in greater detail, with their thicknesses, variations and general dip, as made out from his own observations.

The place of the Venango County Oil-Sand rocks in the series is also shown on the section, and the relation to them of the oil obtained recently from the Beattie well. It is evident that no connection exists between them.

This section will serve another useful purpose. It must impress the reader with a conviction of the *undisturbed condition* of all this part of Pennsylvania. Yet the most productive oil fields of Venango, Clarion, Armstrong and Butler counties are *equally undisturbed.* The popular notion that petroleum wells are dependent upon anticlinals, faults, or other disturbances, is a pure fancy of the imagination. And it is very desirable to get rid of it. Even geologists of standing and reputation are still more or less affected by it; and it is made use of to explain the

---

* On the authority of Mr. Winchell, according to Mr. Randall. There has been no specific determination of these forms as yet.

ascent of the oil from the great "black slate" (Genessee, Huron, Lower Devonian, No. VIII) formation.

On page 70 of the Report of the Geological Survey of Ohio, Vol. 1, Geology, 1873, we read:

"The Huron shale is here [along the shore of Lake Erie] somewhat interstratified with bands of more earthy shale, but exhibits a prevailing black color, and contains nearly 10 per cent of combustible matter. The line of outcrop is everywhere marked by oil and gas springs, and in my judgment, this is the source of the petroleum obtained from the overlying shales and sandstones in Western Pennsylvania. The disturbances which the rocks of that district have suffered, seem to have favoured the disengagement of the oil which emanates from the bituminous shale by spontaneous distillation, while the sandstone strata have afforded convenient reservoirs for its reception."

No words could convey a greater mistake. The district of greatest oil production in Pennsylvania is precisely the district where there never has been any disturbances whatever; and Mr. Randall's section helps very much to make this patent to the least educated eye. And it is only necessary to compare it with Mr. Carll's four Diagrams No. 2, 3, 4, 5, in this volume,—with the pipe line section of Mr. Lucas,—and with the Slippery Rock Section, published with Mr. Wrigley's Report, to see how the same is true of the whole Oil Region. In fact, had not Western Pennsylvania been lifted from the ocean bed into the air (at the end of the Coal Era) in a slow, steady and gentle manner, *without disturbance*, the oil production would never have been a historical event. Western Pennsylvania would have let its petroleum escape into the atmosphere long ago; just as the *highly disturbed* Middle and Eastern portions of the State suffered long ago the loss of all their petroleum by means of their anticlinals and faults, and the universal fracturing of their underground.—J. P. L.]

III.

NOTES ON THE

COMPARATIVE GEOLOGY OF NORTH-EASTERN OHIO, NORTH-WESTERN PENNSYLVANIA AND WESTERN NEW YORK.

By J. P. LESLEY.

CHAPTER I.

The Nomenclature of our rocks.

The selection of proper names in geology is vital to its clear comprehension.

No special names were given to the individual rocks underlying the Coal Measures, in Pennsylvania, by the First Survey. The formations were numbered from I up to XII; but the several sand rocks, shale beds, limestone layers, &c. composing each formation were neither numbered nor named.

Prof. H. D. Rogers, when he published his final report in 1858, rejected the old numbering and gave names to the Formations. But these names have not been adopted by geologists. Only three or four of them can be said to have come even partially into use. Of these the *Seral Conglomerate* (XII), the *Umbral Red Shale* (XI) lying just under it, and the *Vespertine White Sandstone* (X) next below the Umbral, are the best known in Western Pennsylvania; the *Primal Sandstone and Shales* (I) at the bottom of the series, and the *Auroral limestone* (II) next above, are somewhat used in Eastern Pennsylvania.*

---

*It is my intention to discuss the question of Nomenclature in a separate chapter of my general report of the progress of the survey in 1874, under the head of A history of Geological Explorations in Pennsylvania. It is only needful here to state the reasons which seem to make it proper and desirable to follow the universal practise, in colleges and text-books, of using the New York names of the formations wherever they apply, only adding to the list such other names as geologists in other States have found it absolutely necessary to invent.

The New York Geologists on the other hand gave names both to the larger formations and to the smaller groups and series of beds composing them, thus:—

Carboniferous,	Coal measures,		} XII.
	Conglomerate,		
	Carboniferous shales,		XI.
Devonian,	Catskill,	{ Oneonta Sandstone,*	X.
		Montrose Sandstone,*	IX.
	Chemung,		
	Portage,		
	Hamilton,	{ Genesee, Hamilton, Marcellus,	} VIII.
	Upper Helderburg,	{ Corniferous, Scoharie, Caudagalli,	
	Oriskany,		VII.
Upper Silurian,	Lower Helderburg,		} VI.
	Onondaga,		
	Niagara,		
	Clinton,		} V.
	Medina,		
	Oneida,		IV.
Lower Silurian,	Hudson River,		} III.
	Utica,		
	Black River,		} II.
	Birdseye,		
	Chazy,		
	Calciferous,		
	Potsdam,		I.

This example has been universally followed by other State Surveys. Most of the names bestowed have been geographical. The place where a formation was found most largely developed, or the place where a group of beds was first studied or best described, suggested the name. The place might be a remote and insignificant village like Potsdam on the Canada line; or an important centre of (salt) manufacture like Onondaga; or a place

---

* Of Vanuxum, who accepted Catskill from Mather, which turned out to be an unfortunate substitution.

of resort for tourists, like Niagara (falls) and Trenton (falls;) or a mountain range, like the Catskill; or a cardinal point in engineering, like Portage, where the excavations for a series of canal locks furnished an extraordinary number of fossils to the geological museum of the State. Whatever happened to be its origin the name, once accepted by the New York geologists, became fixed in text-books, and followed the outcrops of its rocks into other States. Had the Pennsylvania and Virginia Surveys acquiesced in the New York nomenclature it would have been enlarged and modified southward and westward so as to apply everywhere; and we should now have a consistent and harmonious language for the geology of the United States.

In Pennsylvania the New York system of names could have been practically modified for the use of students in Ohio and Virginia. This was not done. A gap was left between New York and the Middle and Western States. These made their own names for their own geology. The new names thus manufactured were, however, largely geographical; but the places from which the names were taken belong to the several States. In the geology of Tennessee, for example, we have "Ocoee" and "Chilhowee" for *Potsdam*, &c.; "Knox" for *Calciferous*, &c. In Ohio we have "Cincinnati" for *Hudson River*; "Waverly" for *Catskill*, &c. In the west we have Warsaw, Kaskaskia, St. Louis and other places giving names to formations.

Had Pennsylvania added new names of its own to those adopted in New York, to express rocks present in Pennsylvania but absent in New York, Ohio would have adopted these Pennsylvania names for the same rocks passing over into Ohio. In default of such, Ohio has had to give to her own rocks her own geographical names. Now, Pennsylvania must adopt these Ohio names, having none of her own to use in their stead. It would be absurd to give a new name to an already well named rock.

North-western Pennsylvania and North-eastern Ohio have the same rocks.

The Lower Productive Coal Measures, and the formations under them holding Oil, pass from one State to the other; and some of them continue through into Western New York. In other words, Lake Erie is bordered on the east and south, at Dunkirk, at Erie, at Cleveland, by a highland composed of the slightly

uplifted edges of one and the same pile of formations; the outcrops sweeping through North-East Ohio, North-West Pennsylvania, and South-West New York. These outcrops have been
carefully studied by the Ohio geologists, and we shall adopt such
of the local Ohio names as represent rocks not named by the
New York Geologists.

XII Conglomerate. Base of the productive coal measures.

XI Cuyahoga Shale. So named at Cleveland.

X Berea Grit. (First (and Second ?) Mountain Sand.)

VIII { Bedford Shale, (First Mecca Oil-horizon of Read.)
Cleveland Shale, (Chemung.)
Erie Shale, (Chemung and Portage.)
Huron Shale, (Hamilton,) (Second Mecca Oil of Read·)

We shall use, however, in this report chiefly the four special
Ohio names *Cuyahoga* shale, *Berea* grit, *Bedford* shale and *Cleveland* shale, because no names were ever given to these individ-
ualized rocks either in New York or Pennsylvania, and because
in both States, but especially in Pennsylvania, the Berea grit has
been confounded with the Conglomerate No. XII.

The Berea grit is a single formation in Ohio. It begins to be
a double formation (a few feet of shales between two sandstones)
just as it is leaving Ohio to enter Pennsylvania. Before reaching Venango county its two members seem to get widely separated, and to form the *First and Second Mountain-Sands* of Mr.
Carll's sections. The Third Mountain-Sand is not recognized,
nor named, in Ohio. The First, Second, Third, Stray, Fourth
and other Oil-Sands, also, are neither named nor recognized in
Ohio. Just so the great Catskill Formation (No. IX) is not acknowledged in Ohio, and, of course, has no name. Its place would
be below the Berea grit. It is not yet recognized even in Venango and Warren counties, although it is well known on the
Sinnemahoning. Many other Pennsylvania rocks are wanting in
Ohio.

The Ohio set of names is, therefore, not a complete and sufficient one for Pennsylvania. But its *Sub-carboniferous* names
may be adopted by us; just as our Coal-Measure names: "Mahoning Sandstone," "Pittsburg Coal bed," &c., have been adopted
in Ohio. The geology of Warren, Venango, Erie, Crawford and
Beaver counties on our side of the line, is illustrated by Mr. M.

C. Read's reports (of 1873) on Ashtabula, Trumbull, Lake and Geauga counties, and by Dr. Newberry's reports on Cuyahoga and Summit counties, in Ohio. The following summary of these reports will be a useful introduction to these notes on the geology of North-Western Pennsylvania.

CHAPTER II.

General Geology of North-Eastern Ohio, compared with that of North-Western Pennsylvania.

Following the shore of Lake Erie from west to east, the counties under consideration occupy positions represented by the following scheme:

			Erie	*Chautauqua.*
	Lake,	Ashtabula	Crawford	Warren.
Cuyahoga,	Geauga,	Trumbull	Mercer	Venango.
Summit,	Portage,	Mahoning	Lawrence	Butler.

The surface of Summit and Geauga, of N. W. and N. E. Portage, and of S. Trumbull is covered with flat layers of Conglomerate No. XII (Seral.)   Outside of its outcrop runs the bench made by the outcrop of the Berea Grit No. X (Vespertine.) Trumbull has the greater part of its surface made by the Cuyahoga shales (XI) above the Berea, and by the Bedford and Cleveland shales below the Berea; the lowest (block) Coal bed, No. 1, occupying its S. E. Corner and spreading over Mahoning. Most of Ashtabula, and Lake, and all the northern third of Cuyahoga, are occupied by Erie shale (VIII, Chemung and Portage,) which also underlies the bed of the Lake.

These rocks dip very gently south. The Conglomerate forms the highland, 700'—800' above the lake level, and this upland is drained by the Mahoning river into the Ohio. Numerous short and rapid streams cut deep gully-like valleys through the Berea Grit and its underlying shales, northward to the lake.

The same topographical features characterize the geology in Pennsylvania. The branches of the Beaver river and French Creek head up through Mercer and Crawford into Erie, and have their springs within ten miles, and sometimes within five miles, of the lake shore, and at an elevation of five hundred feet and upwards above it. A number of short rapid streams, the Ashtabula, Conneaut, Elk, Walnut, Six Mile and other creeks, cut deep gorges in the opposite direction descending to the lake shore.

The steep slope facing the lake continues eastward through Chautauqua county in New York, and at a much higher elevation. Down its long southern slope descend the branches of the Allegheny river, French Creek, Oil Creek, the Brokenstraw and Conowango creek, at the head of which lies Chatauqua Lake, with its surrounding rock escarpments of Berea Grit, once supposed to be outcrops of the Conglomerate No. XII.

The real outcrop of the Conglomerate No. XII enters Pennsylvania in the extreme north-west corner of Mercer county, and runs across Crawford, south of Meadville, toward the mouth of Oil Creek. East of Oil Creek it is washed away, and its true outcrop is to be sought on the Venango hills south of the Allegheny river, whence it has not been properly traced through Forest into M'Kean. It has been confounded with the Berea Grit (1st or 2d mountain sands) all through the Oil Creek and upper Allegheny river regions. This has caused it to be unduly separated from the great sandrock of the Tionesta Creek region. The work of 1874 has already furnished a clue out of this labyrinth of errors; and it is to be hoped that the surveys of 1875 will rectify this confused part of our geology.

The *Upper* Berea Grit (1st Mountain Sand?) brings its outcrop into Pennsylvania exactly in the south-west corner of Crawford; and the outcrop of the Lower Berea (2d Mountain Sand?) lies a little further to the north. The two outcrops range away past Meadville to Lake Chautauqua, where the Lower (?) Berea makes the great cliffs of the Rock City; and on to Olean, where it makes the other well-known Rock City. The geological map of New York shows it in spots as far north as Ellicottville.

The mountain plain of north-west Crawford, and central and eastern Erie, is composed of the underlying Bedford and Cleveland shales; and the steep lake slope, of Erie shales. All these form part of No. VIII, the Chemung and Portage of New York. Mr. Sherwood maintains that he has traced the Rock City Conglomerate (Lower? Berea Grit; Second Mountain Sand of Mr. Carll) all the way east to the Tioga river, and has proved it to be the top layer of the Chemung, immediately underlying the Old Red (Catskill, No. IX.) If this be so, then we must divide the

Berea Grit formation of Ohio into two distinct formations of very
different ages, and not only insert into its middle the thick shale
formation separating the First Mountain Sand of Mr. Carll from
the Second Mountain Sand, but we must consider this to be a knife
edge of Catskill No. IX.   By reference to Mr. Carll's report, it
may be seen what progress we have made in the direction of
such a conclusion.

CHAPTER III.

The Conglomerate in Ohio.   XII.

In Cuyahoga county, the Conglomerate outcrop at Royalton,
south of Cleveland, approaches within twelve miles of the lake
shore, and in Orange township, within eight miles.   Its base is
less than 500 feet above lake-level.

In Lake county, at Little Mountain, 8 miles from the lake shore,
its base is 700 feet above lake-level, and the top of the cliff
which it forms 750'.   This gives a south dip of 10' to the mile.

Along the east line of Geauga and Summit counties the outcrop of XII is set back* to a distance of thirty-five miles.   Its
outcrop then turns at a right angle due east along the south
line of Trumbull, and northwards along the east line of Trumbull,
and enters Mercer county, Pennsylvania, at West Salem P. O.
Along this back line (the south line of Trumbull) its elevation is
only from 300 to 350 feet above lake-level.   The dip to the south
is, therefore, here at the rate of $\frac{700-350}{35}=10'$ per mile.

In Geauga county (S. E. corner) XII reaches its maximum
thickness in Ohio, viz.: 175 feet.   This is about twenty miles due
south of Little Mountain, Lake county, where it is little over
50 feet thick.

Variations in thickness and coarseness characterize this sand
and gravel formation everywhere.   On the Beaver, Conequenes-

* This is caused by a local W. dip of 4° to 5°.   The Conglomerate once covering Trumbull and Ashtabula lay, therefore, at an ever increasing height
above lake-level going east, and was, therefore, eroded, leaving the Chemung rocks in possession of the country.   There is evidently a N. and S.
anticlinal wave running through the west half of Ashtabula and the west
half of Trumbull counties.

sing and Slippery Rock, in Lawrence county, Pennsylvania, it makes cliffs of 150 feet, and in a few miles thins rapidly away to sixty, twenty-five, and then disappears entirely. On a broader scale the variations are seen in Middle and Eastern Pennsylvania; for the same rock which is 100 feet at Towanda, and 200 at Gallitzin, is 1,200 at Pottsville.*

In Cuyahoga county, Ohio, XII is a coarse sandstone. But in places † some of its layers, especially its bottom layers, are almost (nine-tenths) made up of quartz-pebbles.

In Summit county, XII is quarried in all the townships north of Akron; forms cliffs 100′ high, near the falls in the Cuyahoga Valley; is about 100′ thick; is generally a coarse sandstone, light drab; is sometimes *at its base* a solid mass of quartz-pebbles slightly cemented together; its middle beds are sometimes red or brown, a beautiful building stone,‡ strong and durable.§ Its fossils are broken, floated, macerated fragments of large woody plants,‖ so abundant that their sandstone-casts often make up a large part of the rock, having evidently been piled together here and there like driftwood on a seabeach, all the smaller and more delicate ones having perished without trace. The bark occasionally remains as a thin sheet of coal; in some localities the bark of all of them has become coal; rarely, enough to make an irregular coal bed a few inches thick and a few rods in extent, and *never* with a bed of fire-clay underneath; showing that the plants did not originally grow where they now lie. ¶

---

* Variations in the thickness and coarseness of the Berea Grit, and of all the other *Sand* formations of Western Pennsylvania, have been the true cause of the confusion in our local geology of that region.

† Road from Solon Railroad Station to Chagrin Falls;—near Plank road station W. side of Chagrin Valley.

‡ See the Cuyahoga Falls quarries,—also the very *local* purple rock at Akron, Wolf's quarry, used extensively in Cleveland.

§ This reminds us that the Tionesta (XII?) Sand rock of Forest county, is one of the most perfect building stones in Pennsylvania.

‖ Lepidodendron, Sigillaria, Calamites, and their nuts (Trigonocarpa;) Calamites most common; sometimes entire, with roots. Sometimes the nuts are found *inside* the hollow stems of the Calamites, into which they have been floated.

¶ Impressions have been found on some of the quartz-pebbles; nor can this be considered incredible by those who have seen fac similes of delicate leaves printed at Vienna from copper plates, into the surface of which they had been powerfully pressed. Glass can be made to receive such imprints. Dr. Newberry's explanation is, that the potash of the plant has had a chemical action upon the impressed side of the pebble.

The coarsest parts of XII hold pebbles as large as the fist; commonly their sizes range from egg to nut; all are not rolled fragments of the quartz veins once penetrating the Canadian mountains; there is an intermixture of rolled quartz-slate. Other softer rocks could not furnish pebbles to this conglomerate, because they have been reduced to fine sand, or mud, by the transporting agent, whether water or ice. In Western Pennsylvania the pebbles are all small. But on the upper West Branch Susquehanna there are places where they attain the size of ostrich eggs. In Eastern Pennsylvania they are commonly as large as hen's eggs or nuts; on Broad Top, in Huntingdon county, they are seldom larger than peas.

In Geauga county, Ohio, XII differs greatly at different localities, sometimes worthless, sometimes an excellent building stone. In Russell T. it is a fine-grained, hard, clear white building stone; in Chester it shows 30'–50' ledges of pure quartz-pebbles (with loosely cementing sand) good for glass manufacture.*   In Newberry township in the same county it is handsomely colored. In Parkman township, where it is 175' thick, good building stone layers alternate with pebble rock. In Thompson township, its cliffs and caves attract the lovers of picturesque scenery.

In Trumbull county, Mr. Read says that XII grows thin and sometimes disappears altogether as it approaches Pennsylvania. Some of his sections of the Block Coal Bed show merely patches of conglomerate, constituting sometimes the floor, and sometimes the roof of the bed. In these he recognizes the Conglomerate XII. But Mr. Hodge in 1838 showed that the Sharon Coal (which is the Block Coal No. 1, of Trumbull county, Ohio,) *underlies* the Conglomerate. It is barely possible that Mr. Read has lost his hold on the great 175 foot Conglomerate of Geauga county in coming round through the South of Trumbull county, and could not see it in the 40 foot sandrock *above* the Block Coal because he looked upon this Block Coal as the lowest bed of the Productive Coal Measures, and therefore *over* the Conglomerate. This led him to express an opinion that the Conglomerate XII, *does not extend under the Pennsylvania Coal Field towards Pitts-*

---

* As the cliffs wear away the fallen pebbles and sand are re-cemented into a still harder rock containing modern plants, and perhaps the relics of man.

5—I.

burg.   Whereas cliffs of it sometimes 100 feet high border all the valleys of the Beaver river *

It is impossible to harmonize this part of the geology of Trumbull county with that of the adjoining county across the State line.   Mr. Read's assertion is positive, that everywhere over the Conglomerate (XII) there is in the series a sandrock from which it may be distinguished by several features.   This sandrock overlies the Block Coal No. 1.   *When this coal bed and its shales are absent the sandrock lies immediately on the Conglomerate.*   The sandrock is finer grained, more micaceous; when it has gravel in it the pebbles are not round and are more firmly cemented. The pebbles of XII are very round and loosely cemented; its outcrop is usually less rugged than that of the sandrock above it; and it is more broken by long vertical fissures.   And he offers these features as a practical guide in coal hunting, saying that no coal bed need be looked for under XII or to the north of its outcrop.

Yet it is quite certain that in all the country east of the Pennsylvania-Ohio State line the Sharon Block Coal bed *underlies* the Conglomerate XII, and is represented far and wide by a thin coal bed, and sometimes a little group of beds, whose outcrops extend far to the north of the northernmost patches of the Conglomerate.   In fact, throughout Northern, Middle and Western Pennsylvania, some kind of coal bed can always be found *under* the great Conglomerate.   It is equally true that the Sharon Coal is the Block Coal of Trumbull county, Ohio.   These difficulties can only be cleared up by a careful *instrumental* study of the belt of country occupied by the Conglomerate, the Block Coal Shales and the Berea Grit.

It may, perhaps, appear that Mr. Read's Conglomerate is Mr. Carll's First Mountain Sand, and that the Berea Grit should be wholly confined to the Second Mountain Sand, with which it is already chiefly identified.

---

* He is, however, doubtless right in saying that the pebble rock struck in the Oil Wells of Tuscarawas county, Ohio, cannot be XII, but must be Berea Grit, "which, over a large part of the centre of the State (Ohio) is a true conglomerate."

CHAPTER IV.

The Cuyahoga Shale in Ohio. XI.

Dr. Newberry gave this name to a shale formation underlying the Conglomerate and overlying the Berea Grit, because its two outcrops form the two sides of the valley of the Cuyahoga River for many miles above Cleveland. Its destruction has helped largely to make the top covering of clay,* which lies like a sheet over the country bordering on this part of the lake.

Dr. Newberry makes the Cuyahoga Shale to be the uppermost member of the Waverly Group,† consisting of Cuyahoga Shale, Berea Grit, Bedford Shale and Cleveland Shale ; that is, all from the Conglomerate (XII) down to the Erie Shale (VIII, Chemung.) The geology of Pennsylvania opposes the adoption of this name, for reasons already alluded to, and to be fully stated hereafter. The imperfect series of rocks at Cleveland affords no opportunity for so important a classification. But the name Cuyahoga Shale will stand; for it designates a formation extending eastward, with an ever growing thickness, until on the Schuylkill River it becomes 3000′ thick. In Venango county and all through the oil region, it is a most important set of beds, 200 feet thick, yet without a name, (unless we adopt Prof. Rogers' nomenclature, according to which he calls it Umbral,) or remain satisfied with calling it, as we did forty years ago, No. XI.

In Cuyahoga county, Ohio, XI is 150′–200′ thick, with 100′ of XII above it and 70′ of X (Berea) below it. It is a mass of gray clay shale, with thin flags of fine sandstone here and there between the top and bottom ; valueless for minerals ; making a cold tough soil ; usually barren of fossils ; but in certain places *full* of small shells (*Lingula melia* and *Discina Newberryi*) in its lowest layers just over the Berea grit ;‡ also a few fish scales (*Palæoniscus*) and shark's teeth (*Cladodus*.§)

---

* The Erie Clay of Sir W. E. Logan's Canada Reports.

† His Sub-carboniferous System.

‡ As at Berea and Chagrin falls.

§ It is of importance to note that Mr. Carll's horizon of fish spines is at or above the top of his First Mountain Sand, and always in fragments of thin layers of *red sandstone*.

The beautiful scenery at Cuyahoga Falls is produced by the river cutting through the Conglomerate and then excavating a gorge 300′ deep and a circular plain out of the Cuyahoga Shales. The name for the formation is, therefore, well chosen, having been adopted from this distinguished locality. In two miles the river falls 220′, the thin fine flags of the shale formation forming a series of cascades. Here Dr. Newberry has made one of his admirable studies of the ancient and now more or less filled up valleys which make so striking a feature of our inland geology.

The thin sheets of fine flag stone in this formation are covered with wave-marks, and sometimes with prints of sea-weeds. At the Big Falls a number of these layers, each from 6″ to 1′ thick, lie close above each other for 20′, and make quite a sandstone formation for about a mile; of compact rock, like the East Cleveland "blue stone."*

Cone-in-cone, (German *tutenmergel*) is plentiful near the top of the shales, and points to lime in the mud rock.†

Dr. Newberry finds at Cuyahoga Falls nodules of iron ore (clay iron stone, carbonate of iron) in the top shales of this formation; nodules coated with cone-in-cone. This again corresponds with the Venango county iron ore of Mr. Carll's section.‡

The upper part of the Cuyahoga Shale (XI) is famously fossiliferous at Richfield;§ the list of its species being very large, including some fine crinoids   The same is true of it in Venango county; and in southern Pennsylvania fossils in the limestone of XI are specially abundant.

In Lake county IX is about 180′ thick.

In Geauga county its outcrop is a belt of cold clay soil, covered with a forest of gigantic elms, and making good pasture lands when cleared.

---

* In the Bedford Shale VIII.

† In Venango county, Pa., layers of real limestone occupy the same position; and in Southern Pennsylvania these limestones of XI become massive; they increase in thickness going south; in Virginia and Kentucky a great limestone formation underlies the conglomerate.

‡ And with the well known iron ore bed of northern, middle and eastern Pennsylvania, the ore of XI, just under the Conglomerate, mined at Ralston, and opened at Blossburg, Towanda, all along the face of the Allegheny mountains, around Broad Top, and on the summits of Laurel Hill and Chestnut Ridge in Somerset, Fayette and Westmoreland counties.

§ Also at Medina, Weymouth, Bagdad, &c.

In Trumbull county appear again the flagstone layers in XI, thick enough to quarry for building.*   They pave the streets of Warren.   At the bottom of XI *lingulæ* are abundant, and Newberry's fine fish spine (*Ctenacanthus formosa*) was found. Several have recently been found by Mr. Carll. † *Discinæ*, and whorled shells are numerous in Trumbull county, Ohio, as in Venango county, Pennsylvania.

In Ashtabula county, Wayne township, XI makes a more sandy and gravelly soil than further west.   It passes out of Ohio from Kinsman township, Trumbull county, into the North-west corner of Mercer county, Pennsylvania, and seems to cause the belt of swamps and ponds south of Meadville.

CHAPTER V.

The Berea Grit of Ohio, X.

In Cuyahoga county there are famous quarries of this fine building stone at Berea, where the Lake Shore and C. C. C. railroads cross Rocky River at its falls.   It is there sixty feet thick; a fine, homogenious grindstone, rarely pebbly; further south it becomes a coarse conglomerate ; color, grey, some layers nearly white; drab or light buff at Chagrin Falls and Amherst; difference due to local infiltration, washing out, and peroxidation, variously combined.

Oil men have learned that the color of oil-sand-gravel has something to do with the quantity of oil and water which they are likely to get.   The coloring of rocks has not been studied exhaustively; but enough is known to connect it directly with these principal circumstances: the clayey or sandy character of the rocks above, the quantity of iron and manganese present, the porosity of the rock to be studied, the number of feet or fathoms of rocks between it and the surface, the tightness or looseness of the rocks beneath it, preventing or facilitating its drainage, the abundance or rarity of vertical fissures in it, and the nearness or distance of the outcrop affording exit to its waters.

---

* First Mountain Sand ?

† Those in the cabinet of Mr. Randall, at Warren, Pa., seem to have come from a lower horizon.

The Berea grit, as a general rule in Cuyahoga county, has a more shaly *upper* member of 20', and a more sandy, sometimes massive *lower* member of 40'.*

A *middle* member—a stratum of shale—is sometimes interposed as at Chagrin Falls and Bedford. This must be kept in mind.

The Berea grit contains many fish scales (*Palæoniscus Brainerdi;*) imperfect bones of larger fish; a few shark-teeth (*Cladodus;*) and a large tongue-shaped shell (*Lingula Scotica?*)—Also the long leaflets of a coal-measure plant arranged like stars (*Annularia longifolia.*)

At Berea 500 quarrymen are employed, and $500,000 worth of stone got out: Flags, worth eight cents per square foot; "clear rock," thirty cents per cubic foot; grindstones, $12 to $15 per ton, competing in New England with the Nova Scotia stones. The Independence quarries yield a coarser, warm tinted stone, and the largest grindstones. The fine flagstone quarries at Chagrin Falls are in the *top* layers.

In Summit county the quarries at Peninsula and other points of the Cuyahoga valley, the base of the 60' Berea Grit lies from 30' to 60' above the canal. Some layers are nearly white, and large quantities have been used in the public buildings of Cleveland, Detroit, Buffalo, Oswego, &c., as it is more firm and durable, although harder to work and less even in texture, than that from the Amherst quarries. In 1871, 2,800 car loads left the Peninsula quarries. The active industry of this valley prefigures what is destined to be an important industry in the Tionesta and other branch valleys of the Allegheny river in Pennsylvania.

In Geauga county 40' of Berea Grit lies 180' below the Conglomerate, its outcrop ranging southward, outside of (east of) and parallel to that of the Conglomerate, the whole dipping *westward*. It is massive; good for building stone; † and at one spot (Big Gull in Chardon) will yield grindstones equal to the best from the Berea quarries. The Thompson "ledges" yield capital flagging of any desirable size, 8″ to 10″ thick. Two sections are given by Mr. Read to illustrate variations in the formation :—

---

* At the Brandywine falls, Summit county, this formation is in one solid stratum.

† The Chardon court house is from a quarry at Munson.

Upper part, eroded off, · · · · ·   ?′

Sandstone; compact, two layers, · · ·   8′

"   shaly; cleavage horizontal, · - 7′ to 8′

"   "   "   oblique and curved,   6′

"   "   "   horizontal, · ·   4′

"   "   "   curved, - · ·   8′

$$?′ + 34′$$

Sandstone, shaly, in thin layers, - · ·   8′ —10′

Shale, blue, · · · · · · $\frac{1}{2}$′ — 1′

Sandstone, in 8″ to 24″ layers, - · ·   6′ — 8′

"   massive, in two layers, · ·   10′ —12′

"   in thin layers, - · · ·   8′ —10′

$$32\tfrac{1}{2}′—41′$$

Other sections repeat these variations in some different order; and others again show all the layers massive; the changes are rapid, and the oblique lamination quickly becomes horizontal or vice versa, so that great care has to be exercised in selecting a place to quarry.

In Ashtabula county the Berea Grit varies from 40′ to 60′, and from thin to massive layers; generally gray from minute specks of iron, and sometimes iron stained. At Footville, a few of its layers are quarried for whetstones. Its outcrop is marked by a bench covered with sandstone fragments and a thin sheet of soil. It is a little quarried at Winsor mills, where fine exposures show in the gorge, and promise a large future trade. Hard, strong layers of Berea Grit crop out in Colebrooke and Wayne townships, but too much stained for architectural purposes. From Williamsfield its outcrop, now double, passes into Pennsylvania.

In Trumbull county the Berea Grit has been long quarried at Mesopotamia for scythe stones, and some layers would yield fine grit grindstones. In Southampton, Champion and Mecca townships, the formation is deeply covered with drift clay; and in Johnston, Gustavus and Wayne, on the other (east) side of the Musquito Creek valley, as deeply covered with drift sand; but it is struck in all the Mecca oil-wells, just as our First Mountain Sand (=Berea grit) is struck first in all the wells of Venango county around Pleasantville.

It outcrops north and south along the valley of the Pymatuning, both sides, and shows at Vernon massive layers, from which strong firm building stone of any size can be got; but holding iron ore balls. At the Pennsylvania line in Vernon it is full of water-worn quartz pebbles, and has been mistaken for the Conglomerate (XII) which caps the ridge a hundred feet higher up.

The Berea Grit leaves Ohio and enters Pennsylvania where Trumbull, (O.) Ashtabula, (O.) Crawford (Pa.) and Mercer (Pa.) counties corner on each other. But it is no longer, even in appearance, a single sandstone formation. Two Berea Grits outcrop in the ravines south-east of West Williamsfield, the *lower* separated from the *upper* by 15′ to 20′ of clay-shales. This Lower Berea Grit "is apparently thin, and gives promise of no really first class stone." We have seen (above) that this lower member of the formation is thick and massive in the counties towards Cleveland. We will perhaps be able to prove by our surveys of 1875 that this lower member becomes the great Second Mountain-Sand of Venango county, and the conglomerate of Lake Chatauqua; of Olean; of Darling, west of Warren, Pa.

The Berea Grit of the Trumbull county ridges, east and west of Musquito creek, was penetrated by the Mecca oil-wells, and was found to be soft, porous and *charged with petroleum.*

The Mecca oil came from this Berea Grit (First and Second Mountain-Sand of Venango) and from the Bedford Shales next below it.

The wells sunk on both ridges yielded oil; but those on the *east* ridge, where the Berea Grit was covered with Drift sand instead of Drift clay, yielded little; those on the *west* ridge where the slow escape of the oil through ages has been retarded by a thick deposit of Drift clay yielded more; but all were soon exhausted. They are recuperating, however; and the best may yield again in paying quantities.

CHAPTER VI.

The Bedford Shale in Ohio. IX?

Red Shale,—a bright red shale formation *under*lies the Berea Grit in most parts of Cuyahoga county, and serves as a guide to the quarrymen. It *over*lies the "Bluestone" of the East Cleveland* quarrymen.

Instead of red shale, sometimes blue shale is seen, as in the gorge of Tinker's Creek, in Bedford, 70′ thick. The difference is due to the quantity and oxidation of iron contained in the clay. The "Bluestone" is the same bottom part of the Bedford shale in the shape of a fine grained sandstone, containing a good deal of iron pyrites, which turns red on long exposure. It saws up into fine flagging.

The Buena Vista stone of the Scioto Valley used for the streets of Cincinnati and New York is the same rock; and is no better.

At Bedford the bottom layers hold multitudes of several species of shells, especially Winchell's large spirifer (*Syringothyris typa*,) found also by Mr. Randall, at Warren, Pennsylvania (?) Also *Rhynconella sagerana, Orthis mitchelini, Spiriferina solidorostris, Macrodon hamiltoniœ*, &c.

In Summit county, below the Brandywine falls, the *Syringothyris* is very conspicuous, and cabinet specimens can be got in the shape of slabs of rock thickly set with them.

In Geauga county the formation is 40′ to 50′ thick in the ravines of Grand and Chagrin rivers. It includes layers of sandstone, compact and fine grained, *susceptible of polish*, 1′ to 3′ thick, good for window caps and sills, but containing iron, and, therefore, liable to stain or "run;" some would do for oilstones.

In Lake county, (Kirkland township,) the Bedford is 40′ thick; mostly *rock*, hard, compact, in layers from 1′ to 18′. To the eastward it is shale again.

---

* At Kingsbury and in several places in Newburg and Bedford; and everywhere to the west of the Cuyahoga.

In Ashtabula county, west side, the Bedford shales are only 35' thick, soft, friable, and making a tough clay soil. Towards the Pennsylvania Line, eastward, they grow thicker, harder and more sandy, and include thin layers of fine sandstone, but no quarry-rock.

In Trumbull county, also, the Bedford is mostly concealed beneath Drift. On the west side of the county (in Mesopotamia) two feet of shales exactly like Bedford shale separate the Upper and Lower Berea Grit. At the Pennsylvania Line (in Williamsfield) these two feet have become 15' to 20' of alternate soft and hard band, holding large *Lingulæ* and other shells, and resembling in all respects the ordinary top layers of the Bedford formation.

Mr. Read's description (page 508) leaves it to be supposed that he considers the *Lower* Berea to be one or more of the thin Sand layers of the middle (or lower part) of the Bedford Shale, growing sandy and massive as it approaches Pennsylvania, and enclosing between itself and the *Upper* Berea, the top portion (with *lingulæ*, &c.,) of the Bedford Shale. In this case we must read our column downwards, thus:

Conglomerate, XII. Seral.
Cuyahoga Shale, XI. Umbral.
Upper Berea Grit, X. Vespertine. First Mountain-Sand.
Upper Bedford Shale, *Red*, IX. Catskill (Red.)
Lower Berea Grit, VIII. (Mr. Sherwood.) Chemung.
Lower Bedford Shale, *Gray*, VIII. Chemung.
Cleveland Shale, *Black*, VIII. Chemung.
Erie Shale, VIII. Chemung and Portage.

The Bedford Shale yielded oil to the wells at Mecca; and Mr. Read designates it in his section (on p. 506 Ohio Report, 1873,) one of the *Mecca oil-rocks;* the Berea Grit being one also. He calls the underlying Cleveland shale the *first oil-producing rock*, and the Huron shale, under the Erie shale, the *second oil-producing rock*. By this he intimates his opinion that the oil has ascended from the Huron and Cleveland to be caught and held in the Bedford and Berea. The objections to such a view are numerous and cogent.

CHAPTER VII.

The Cleveland Shale in Ohio. VIII.

Black bituminous Shales, 20′ to 80′ between Vermillion river and the Pennsylvania Line, but not exceeding 60′ in Cuyahoga county, underlie the red and blue Bedford.

Oil Springs mark its outcrop, especially in East Cleveland, Grafton and Liverpool, and Mecca in Trumbull county. The shale contains ten to fifteen per cent. of combustible material.*

It is not entirely destitute of fossils. At Newburg Falls "scarcely a fragment of it can be found which does not contain fish-scales." At Bedford shark-teeth are found in it, (*Polyrhizodus*, *Cadodus*, *Orodus*, all coal-measure sharks.) The surfaces of the shale are also covered with little bony cones of the skin of sturgeon-like fish (Conodonts.)

In Summit county the Cleveland Shale is about 50′ thick, without fossils. Oil springs and gas springs from these shales along the Cuyahoga Valley have tempted people to bore for oil in the bottoms.

In Geauga county the Cleveland Shale is only 40′ thick.

In Lake county there are 30′ of black bituminous Cleveland Shale; under which lie 35′ of transition shales, overlying the Erie Shales.

In Ashtabula county (Trumbull township) the same transition is noticeable. No waterfalls here mark the sudden change from black (Cleveland) shale to soft blue (Erie) shale, as in Cuyahoga county. The black shale smells strongly of petroleum, splits neatly into thin layers, contains a small percentage of iron, and weathers into a stiff clay.

The lower or transition layers contain beautiful shells, (*Discina Newberryi*) and a profusion of *Conularia* fossils, which are found in the Cuyahoga Shale of Trumbull county at Vernon.

---

* One analysis gives earthy matter 87.10; fixed carbon 4.90; volatile carbon 6.90; water 1.10; gas per pound 0.62 cubic feet.

CHAPTER VIII.

The Erie Shale in Ohio. VII.

This name has been given by Dr. Newberry to all the rocks which face the southern shore of Lake Erie, east of Vermillion river, and form the bed of the lake between the shore and a more or less straight line drawn from Vermillion river to Eighteen Mile run, south of Buffalo.

The whole thickness of this formation is 400' to 500' opposite Cleveland ; 1,200' and more at Ashtabula ;* about 2,500' ? at Erie ;† 3,000' and more along the Genesee river in New York, and about 3,200' at Huntingdon and Catawissa in Pennsylvania.

It is evident that the origin of the deposit was in the east. It not only thins away westward but becomes finer and more muddy in the same direction, as will clearly appear in the next chapter, treating of the Portage Formation in New York.

Dr. Newberry's measurements make it plain that it continues to thin away westward from Cleveland.   The bottom line of its outcrop (or the top line of the black slates) comes out of the lake at the mouth of Vermillion river and runs south past Norwalk, Columbus and Chillicothe to the Ohio river below Portsmouth. At Norwalk, twenty miles south of the Lake Shore, the *supposed* top line approaches and joins the bottom line, reducing the thickness of this *Erie* (Portage and Chemung) formation to *nothing*.   The *Huron* (Hamilton) formation, under it, henceforward (southward) supports Newberry's *Waverly* (Cleveland, Bedford, Berea and Cuyahoga) directly and without any intervention of *Erie* (Portage and Chemung) between them.

---

* Mr. P. H. Watson's well at Ashtabula struck the Huron (Genesee-Hamilton) black slates at 845', which marks the bottom of the formation under the lake.   The top of it is found on the hill slope twenty miles to the south of Ashtabula, at a height of 300' above the lake level.

† Rogers' Final Report I, p. 580, says 1,900' above water.   The whole thickness would then be estimated thus: Lowest Coal Rocks 1,300'-1,400' above Lake Erie.   Add 600' for a dip of 14' or 15' per mile for 40 miles between coal rocks and Lake Shore.   Add 500' or 600' for submerged Portage.   Total 2,400'-2,600'.   From this must be deducted whatever, under the Meadville top conglomerate, corresponds to Bedford and Cleveland Shales, viz: 100', according to Mr. Read in Ashtabula county.

But inasmuch as Dr. Newberry finds in this thin western knife-edge of *Erie* such shells as *Spirifer disjunctus, Spirifer altus, Leiorhyncus mesacostalis, Orthis typa*; fossils characteristic only of the Chemung in New York, together with other fossils unknown in New York, he concludes that his *Erie* shale formation is not Portage and Chemung, but Chemung alone, without any Portage. If so, the Portage has its own knife-edge more to the north-east; perhaps under the lake. And if we could trust implicitly the evidence of fossils we might accept this conclusion. But inasmuch as the *quality of this*, as of every other *deposit*, determined the kind of animals which could inhabit it when it was ocean-bed, and inasmuch as the quality of the *Erie* shale in Ohio is that of the Portage shale in New York, and not that of the Chemung, we are left at liberty to suppose either that the western edge of the Chemung really extended over and beyond the western edge of the Portage, or, which is just as probable, that the abovementioned shells lived first in Portage days in Ohio, and afterwards continued to live, in Chemung days, in New York; just as the beautiful little *avicula speciosa* began to inhabit the old Maryland and Virginia ocean in the Upper Hamilton (Genesee black slate) age, and emigrated as far north as New York not until the next Lower-Portage age.*

It is even probable, from Dr. Newberry's description† of the *Erie* shale, that the two knife-edges of the two formations (Portage and Chemung) lie on top of one another along the Vermillion river out-crop; for he says:

West of Cleveland, the Erie shales are seen to form two beds or groups of strata, of which the upper [Chemung] nearly 100 feet thick, consists of green, gray and blue shales, generally

---

*Prof. H. D. Rogers advanced the same view in 1858, (Final Report, Geological, Pa. Vol. I, p. 580,) in these significant words, the true force of which has been, of late years, acknowledged by some of the most distinguished palæontologists of Europe and America :—

"Thus, there can be no universally characteristic fossils; and each species, or small group of species is only locally—if we view the entire surface of the globe—typical of any formation, or of any portion of geological time."

The assertion thus made by Mr. Rogers, as it was made still earlier by James Hall, is accepted now by all experienced geologists, viz: That the geographical extent, as well as the vertical range, of fossils must depend on the quality of the sediments in which they lived, died and were buried.

† On page 163, Ohio Report. Geology I, 1873.

very soft and fine, interstratified with sheets of micaceous* silvery sandstone, from half an inch to two inches in thickness, with flattened lenticular masses of argillaceous iron ore. These thin bands of sandstone are, sometimes, sufficiently thick and firm to be used as flagging.

The lower [Portage?] series consists almost exclusively of blue and green shales, with thin strata of iron ore.

The whole weathers in smooth homogeneous cliffs, of which the prevailing color is a greenish gray.

These two groups are exposed in the cliffs between the Cleveland and Rocky River; the lower group of beds making the cliff for three miles west of Cleveland, and then sinking into the lake with a strong west dip. The upper group forms the whole cliff at Rocky river, and lies horizontal as far as Avon point. Here the dip changes (to east,) and the lower group comes out of the water and forms the shore bluff to and beyond Vermillion river; then the underlying Huron (Hamilton-Genesee) black slates appear.

In Cuyahoga county† the *Erie* (Chemung) shales reach up the cliffs to a height of 100′ to 150′, and wall-in all the ravines leading inland with layers of gray or blue clay-shales with single layers of pearly mica-sand, and flattened clay-iron balls.

Fifty feet below the top occur the Chemung fossils *Leiohryncus‡ mesacostalis, Spirifer Verneuilli, Orthis Tioga* and *Spirifer altus.* The first two, with a *Rhynconella*, are in great abundance 50′ below the top of the Erie; also on the upper waters of Ashtabula and Conneaut creeks further east. All four may be found in Summit county where the Cuyahoga cuts 100′ down into Erie shales.

In Lake county the top of the Erie in all the ravines is 350′ above lake level. It is a pretty uniform mass of mud shales, blue, friable, with an occasional thin layer of sandstone, hard and limey, blocked by vertical seams, and often containing iron ore balls marked on their undersides with the print of sea-weeds. In these balls are sometimes found fossil shells; and among them has been discovered a new crustacean.§

---

* No rock element will float further than finely broken Mica.

† Dr. Newberry's Report of that County.

‡ A new Leiorhyncus has been found at a lower level in the Ashtabula river harbor.

§ Le Roy township, Paine's Creek.

These layers become in places true limestone; and springs issuing from their outcrops deposit tufa.

In Ashtabula county the Erie shale mass keeps the above mentioned character, sometimes weathering red and making a stiff yellow clay. Its hard layers are sometimes a foot thick and form little terraces. All the lake streams flow in narrow gorges 100' deep; the upland is very wet, unless underdrained by the farmer; but the soil is then seen to be fertile, and especially good for grass and orchards.

CHAPTER IX.

The Disappearance of the Erie Shales westward in Ohio, and their distribution under the whole Coal Field.

Dr. Newberry's fortunate discovery of Chemung fossils in the *upper* division of his *Erie* shale formation, has made it very probable that the Chemung rocks extend as far as Norwalk, in Ohio, and there thin away to nothing, as was said in the last chapter, on the authority of his annual report of 1873.

As a general rule, however, (he says) the Erie shale formation is remarkably destitute of fossils. In New York, on the contrary, the Chemung rocks are, according to Hall, copiously supplied with forms of life. Very rich fossil beds have been found in the Chemung of Warren and Venango counties, Pennsylvania.* Whereas the Portage rocks of Western New York are as remarkably destitute of fossils. This was the reason why the Erie shales of the Cleveland lake shore were considered, until Dr. Newberry's discovery, by every one, as of Portage age.

*Mr. Hatch has just discovered (June, 1875,) two localities north-west of Pleasantville, below the Third Mountain Sand, full of Chemung fossils.

One of these two localities lies 3 miles N. 50° W. from Pleasantville; here a thin band of rock outcrops 1,218' above tide, and 300' below the Church Run (Garland) conglomerate (Second Mountain Sand,) and contains an abundance of *Platyceras, Avicula, Aviculopecten, Productus, Rhynconella, Strophomena,* a large *Lingula;* Crinoids; a large fucoid; and perhaps some fish remains.

The other is 2½ miles N. W. of Pleasantville, where an outcrop of a band of fine sandstone 1,263' above tide, and 250' below the same conglomerate, exhibits plenty of *Avicula, Productus, Spirifer, Orthis, Strophomena, Strophodonta, Orthoceras;* some gasteropods, probably *Nautilus* and *Straparolla:* and crinoids. It is a very rich locality.

As Dr. Newberry divides his Érie into an upper fossiliferous member and a lower nonfossiliferous member, it would naturally follow that the lower one is as much Portage as the upper one is Chemung. Both, then, seem to disappear at Norwalk together. It is a knife-edge of the whole Portage-Chemung division of our No. VIII.

The question now arises: Does this knife-edge in Ohio of a mass of sediments which in Middle Pennsylvania are more than 3,000 feet thick betoken the existence of an ancient shore to dry land in Central and Southern Ohio in Devonian days?

Certainly not. For then the tapering would have been sudden; and the deposit would have been coarse sand or gravel.

It must then betoken the extreme distance out to sea,* westward, to which the finer parts of the deposit floated, brought as they were by rivers into the sea at the coast line of a continent lying in the direction of the present Atlantic, or the present New England and Northern New York.

Is it possible to determine the line of this extreme westward floating of Portage and Chemung mud, mica flakes and fine sand particles—covered up as everything now is by the *Waverly* and *Carboniferous* deposits of South-eastern Ohio, Western Pennsylvania and West Virginia?

Certainly not, until the borings of the Oil Region go deep enough to strike the top of the *Huron* (Hamilton-Genesee) black shale, and at points sufficiently numerous to enable us to calculate the rate of thinning (of Portage and Chemung) southwestward; that is, down the Ohio River. All we say at present is, that, as the Portage and Chemung deposits are very thick on the Upper Potomac, and are quoted as 750′ in East Tennessee, the line of disappearance must be supposed to run a little east of south through Ohio, passing under the Ohio river bed somewhere near Pomeroy, but at a depth far beyond that of the deepest salt wells. The Oil Break anticlinal of West Virginia

---

* Saying this of course implies dissent from Dr. Newberry's conclusion (p. 165 Ohio Report Geol. I, 1873,) that the *Erie* Shale "water basin was much more shallow and narrower than that in which the *Huron* shale accumulated," and that the conglomerates of the New York and Pennsylvania Line region are proofs of "oscillations of sea level which sometimes brought shore lines to the margin of Ohio, but never produced any dry land in the eastern part of the State." We are not yet prepared for a conclusion which takes for granted that the oil-gravels were shore deposits.

is so slight a disturbance, and its numerous wells have fallen so far short of the required depth that it throws no light on the subject.*   It only serves to confirm the fact taught by all the Oil borings of Pennsylvania, that it is not proper to refer our Petroleum to a mother rock lying at a still greater depth than the Erie Shale, namely, the *Huron* (Hamilton-Genesee) *black slate* as Mr. Read does in his report on Trumbull county, Ohio Report, 1873, p. 507, where he says:

"The lower Oil-producing rock, marked 8 in the section, [Mecca Oil region section on p. 506 ; *Huron* shale] * * * is here not less than 1,200 feet from the surface, &c.   The great supply of petroleum produced in Pennsylvania is obtained from the deposit marked 8 or its equivalent."   The fact is, that at Mecca, the Portage and Chemung alone are more than 1,200' thick, (and probably much more ;) and they thicken eastward to such a degree that the 1,500' wells of Butler got no lower down in the Chemung than to the Third and Fourth Oil Sand ; leaving an unknown thickness of Lower Chemung still unpenetrated ; and beneath that again, the entire thickness of the Portage, which Hall makes 1,400' at its outcrop on Lake Erie.   So that the bottom of the Portage (or the top of the Huron black shale) must lie well nigh 1,500' beneath the bottom of the deepest oil-wells of Butler county.   Were the Butler county sinkers to touch Huron black slate even at 3,000' beneath the surface they would still have to sink through 700' more of the Hamilton Group (Genesee, Hamilton and Marcellus) before they would reach the top of the Corniferous (Upper Helderburg) Limestone, which is the probable mother rock of the Canada petroleum.†

Mr. Watson's deep well at Titusville, if we had a good record of its measures, would teach much.   Its total depth is quoted by Mr. Wrigley at 2,114 feet.‡

---

* Professor Andrews in his Report of 1870, (Geological Survey, Ohio, 1870, Part II, page 66,) gives a section of a well at Burning Spring, by A. B. M'-Farland, Esq., 1,021 feet deep to 15' or 20' of black slate, which he thinks to be Waverly (Cleveland) black shales.   If it be, the Huron shales must lie at a far greater depth.

† See Reports of the Geological Survey of Canada ; and Articles X and XI of Chemical and Geological Essays by Thos. Sterry Hunt, late Chemist of the Canada Survey.   Boston—Jas. R. Osgood & Co., 1875.

‡ Mouth 1,181' above tide and 426' below Lake Erie.   Bottom 933' below tide and 1,506' below Lake Erie.

Mr. Watson's new well is said to have reached (May, 1875,) 2,700′,* and is reported to have black slate in the bottom. This makes a total vertical distance of about 3,300′† from the Conglomerate (XII) down to the bottom of the new well, and about 4,400′ from XII down to the Canada oil-rocks; which would settle the question of the depth of the Canada Oil horizon beneath the Pennsylvania Oil horizon, making it about 3,250′.

It would also give 17.5′ per mile as the *average general dip* of the top of the *Huron* (or bottom of the *Erie*) from Dunkirk to Titusville. thus :

Tide level.		*Tide level.*
		Top of Portage S. S. (Hall) 1,653′+
1,181′+	Top of Titusville well.	
731′+	Top of 3d Oil S. (Wrigley)	Top of Dunkirk well (Wrigley,) - - - - - 753′+
		Top of Huron (Genesee) slate, 253′+
		Top of (Marcellus?) slate, - 22′—
		Top of Corniferous limestone, 445′—
933′—	Bottom of Watson's old deep well (Wrigley.)	
1,519′—	Bottom of Watson's new well (May 1875.)	

This would enable us to give a tolerably certain answer to the question whether the Portage upper sandstones of New York are the Oil Sands of Venango county, or not? For the Canada oil-rock (Corniferous) is only 445′+1,653′=2,098′ below the Portage Sands; but 3,250′ below our Oil-sands.

CHAPTER X.

The Portage Group in New York. VIII.

The Portage‡ group of New York was subdivided by Mr. Hall, in his surveys of 1835 to 1843, into :—

Upper; *Portage Sandstones ;*
Middle; *Gardeau Shale and flagstones ;* and
Lower; *Cashaqua Shale.*

The shales of Cashaqua creek, lying immediately upon the upper black slate (Genesee) of the Hamilton Group, are 110′ thick on the creek, but only 33′ where they appear on the shore of Lake Erie, in the bluff at the mouth of Eighteen Mile run.

---

* Bottom 1,519′ below tide, and 2,092′ below Lake Erie.
† Taking Mr. Carll's section at Pleasantville as a guide.
‡ Called first the Nunda Group from the town on the Genesee river, afterwards Portage. James Hall's 4th Report of 1843, p. 224.

They are soft, muddy, crumbling, green ; fossils rare but pecu-liar, and traceable, as *characteristic* of this formation, for 150 miles east and west through the State of New York.   Nor are they accompanied or mixed in with fossils belonging to other horizons.*   The formation contains lime concretions, but not in layers.   On Seneca lake the green shales are interbedded with thin flagstones and sandy shale.   The formation, therefore, is growing sandy eastwards ; and further east still, it is all of it thinly laminated sandy shale.   West of the Genesee its lowest layers are blackish and separated from the underlying black Genesee slates by a thin lime band.   It passes through Wyom-ing on its way to the lake.

The Gardeau shale and flagstone series, on the high precipit-ous banks of the Genesee exhibits its bottom layers of green slaty and sandy shale and black slaty shale, with one or two courses of sandstone.   Higher up the sandstone layers become thicker and more numerous, separated by green and black shales. In the uppermost parts of the series the sandstones become massy and the intervening shales also thicker.   East from the Genesee River the whole formation gains in sand, shales growing scarce.   West from the Genesee the sand gradually disappears and the formation turns more and more to shale.   On Lake Erie, *over the Cashaqua shale*, lie thick *black shale ;* over that alter-nations of green and *black shales* for several hundred feet.   On the Genesee fossils are scarce, but on the Lake Shore numerous.

The Portage Sandstones are well exposed in the perpendicu-lar walls of the deep gorge of the Genesee at Portageville, walls 350 feet high ; the upper layers being thick bedded sandstone with little shale ; the lower part thin sand layers with frequent alternations of shale.   This formation was separated from the one below (Gardeau) because of the massiveness of its sandrocks and the presence of *vertical* sea plants in them.   On the thinner Gardeau flagstones the sea weeds lie *flat*.   This attitude of the plants is usually sufficient to determine their places in the series.†

---

* Same, p. 243.—The most common are Avicula speciosa, Ungulina subor-bicularis, Billerophon expansus, Orthoceras aciculum, Clymenia? compla-nata, Goniatites sinuosus, Pinnopsis acutirostra, Pinnopsis ornatus.

† These two kind of seaweed are well exhibited in figures 104 and 105, pp. 241, 242 of Hall's Final Report of 1843.   "In the absence of fossil shells," he says, "We have a great abundance of marine vegetation, or fucoids, and

But the distinction between Portage and Gardeau vanishes towards the east.   Towards the west, where the Gardeau becomes all mud, the Portage becomes an alternation of mud and thin sand rocks.   On Lake Erie it is hard to find the *massive* Portage sandstones of the Genesee gorge.   One such appears at Laona.   Another forms the end bluff at Shumla, and elsewhere along the Lake Shore, especially in the New York and Erie Railroad cut west of Fredonia.

It is evident that the sand came from the east.   The gradual westward decrease of sand and increase of clay, in the whole Portage group, has permitted the excavation of Lake Erie.   Had the Portage group been a mass of mud rock all the way from Ohio to Central New York, Lake Erie would have been extended at least as far as the Genesee River, if not further.

It is specially interesting to the students of Pennsylvania Geology to remark this gradual passage of sandstones into shales going west; for a similar change in the same direction has happened in the Oil Measures under Venango and Crawford counties.

The difference between the Portage Group as a whole and the overlying Chemung Group as a whole is thus stated by Mr. Hall.   The Portage Sands are finer grained and always more clayey than the Chemung Sands.

This observation made along the New York outcrops is confirmed by the Oil borings.   Under the coarse, often gravelly, Oil Sands of the Chemung, are the fine sands and sandy shales

these are very characteristic of the Portage group.   Scarcely a locality can be examined where one or more species does not occur.   The horizontal seaweed *Fucoides graphica*, occur in short stiff fragments, lying in great confusion, *spread over the surface of the thin flagstones of the middle part of the group.* * * * It may be seen on the sidewalks of all the villages upon or near the group; some of the finest * * * in the streets of Geneva; at Penn Yan * * * less perfect.   The beds of the ravines are strewed with them.   This species disappears towards the upper part of the group where the sandstones become thicker and gives place to another species, *Fucoides verticalis;* small round stems, upright, as if growing when the sand was deposited around them.   This species has always been found *characteristic of the upper part of the group.*   Although vertical seaweeds have been found in formations above the Portage flags, they are always of a different type from these.   This Portage fucoid may be seen at the Lower Falls, and in many of the sandstones higher up; but it is most abundant in the Upper Sandstone at Portage, and the uppermost mass of the group is everywhere known by its presence.

of the Portage.   Hence the absence of good oil-rocks in the deepest wells.

Fossils are rare in the Portage Group, except fucoids (sea weeds.)   The *only* Portage bivalve shells (brachiopods) are *Delthyris levis* and *Orthis tenuistriata*, both peculiar to this group. In the Hamilton group below and in the Chemung group above brachiopod shells are ten times more numerous.   The principal Portage life forms are several species of *Goniatites* (*bicostatus* and *sinuosus*) *Bellerophon striatus*, a *Pterinea?* and the little *avicula speciosa*.

In the Chemung rocks Portage fossils are totally wanting, or extremely rare; but the fossils of the Chemung itself are numerous. To this assertion in the report of 1843 is added another, which requires important modifications, viz : That the Portage Massive Sandstone formation may be traced through Erie county, Pennsylvania, into Ohio, where it appears as the Waverly Sandstone, everywhere separating the almost non-fossiliferous shales and sandstones (Portage) below it, from the highly fossiliferous sandstones and shales (Chemung) above it.

If the Waverly sandstone (Berea Grit) be *Mountain*-sand, and the Portage sandstones be below the *Oil*-sands, as we shall see, then Portage cannot be Waverly.

Oblique bedding and ripple marks in the coarse Portage Sandstones show the force of currents and also the shallowness of the water in which they were deposited.

Shrinkage cracks have been seen in relief on the undersurface of Portage clay-sandstone layers resting on soft black shale, and prove that the water was so shallow in some places sometimes, as to leave the bottom exposed to dry in the sunshine.   The blackness of the shale is, no doubt, due to the animal or vegetable sea-life which soaked into this alternately covered and uncovered shore mud.

Concretions are more or less calcareous in all the shales of the Portage group.   Perfectly round ones, seamed with crystalline limy matter, are found in the *black* shales ; flattish oval ones, with less lime, in the *green* shales.   Hundreds of them are annually undermined from the steep banks of the lake shore, and are burnt, to make hydraulic cement, between Dunkirk and Portland harbor, where they are often 2′ to 3′ in diameter, and

6″ thick. Mr. Hall saw one 6′ in diameter, which had fallen from a bed of black shale near Sturgeon point. On the Genesee river many of them have the cone-in-cone structure for an inch or two in from their surface. It is common to find cone-in-cone along the whole lake shore belt in New York, Pennsylvania and Ohio. Beautiful specimens are found near Erie.

Casts of flowing mud are also to be seen on Portage flags, looking like the cooled slag from an iron furnace.* Casts of mud furrows into which multitudes of little shells have swept and been fossilized, are to be found on the underside of Portage sandstone layers ; the agent being, of course, some tidal current, or the currents set up by prevailing winds along the shore of the ancient Portage sea. Sets of long straight parallel scratches are seen over wide areas ; these are sometimes crossed by others at an angle, as if glacial ice had moved over rock, or thin shore-ice had lightly scratched the shallow bottom while still soft. It is remarkable that the general direction of such scratches is nearly east and west. Sometimes one end of the scratch is deep and large, and the other tapering to nothing, as if the object had struck hard and floated off.

The Portage Sands form vertical cliffs like walls of regular masonry at the south end of Cayuga and Seneca lakes, and isolated masses project like pilasters or columns. The lower (Cashaqua) shales are finely shown at the outlet of Crooked lake ; the Upper (Portage) Sands at the head of the lake, covered with *fucoides graphica*. The whole of the Portage Group is exposed at the head of Canandaigua lake and in the ravines around Naples ; and in the sides of Honeoye, Hemlock and Canadice little lakes ; and along Caneseraga Creek near Dansville ; and especially along Cashaqua Creek ; and in the gorge of the Genesee from Portage to Mount Morris ; and further west along Allen's Creek and Tonawana Creek ; and on the Lake Shore from Eighteen Mile run into Pennsylvania, in cliffs from 20′ to 100′ high ; and up the ravines of Twenty Mile Creek, Chautauqua Creek and Canadawa Creek.

In the gorge of Chautauqua Creek may be studied the overlay of the Chemung Group upon the Portage Group.

---

*See plates of this and of cone-in-cone, on pp. 232, 233, 235, 237, of Hall's Report, 4°, 1843.

The whole thickness of Portage (lower, middle and upper) rocks cannot be less than 1,000'.

The Genesee River falls 600' from the head of the Falls to the bottom layers of the group near Mount Morris. Add 200' for the Sandstone cliffs above the top of the Falls and 300' for ten miles of dip, at the average rate of 30' per mile, and we have 1,100 feet.

Again: Thirty miles of lake shore Portage exposures extend between Eighteen Mile run and Chautauqua Creek, where 300'– 400' of Upper Portage flags are exposed. This would make the total thickness of the Portage Group nearly 1,400'.*

---

* The *general average dip*, got from a great number of sections, was fixed by Mr. Hall at 25' per mile; but the larger proportion of *local average dips* gave him 50' to the mile. Numerous slight undulations, however, were found to traverse these sections, which if taken into account would have diminished this rate. These local undulations make it unsafe to use the dip with confidence as a basis of calculation.

CHAPTER XI.

The Chemung Group in New York. VIII.

Highly fossiliferous shales and thin bedded sands, best developed along the Chemung river in middle New York, overlie the fucoidal Upper Portage sandstones (with sea-weed impressions) in the gorge of the Genesee river.

No strongly marked transition from Portage to Chemung can be designated, beyond the fact that a multitude of shells now appear which did not exist in the Portage sediments.   There is also generally a greater coarseness of grain to the sand-rock layers of the Chemung group.

The whole mass of Chemung may be said to consist of alternate deposits of thin flagstones and of shales; with frequent layers of impure limestone or limy shale, resulting from the presence of great colonies of shells.

Sandstones and shales alike weather *on the surface* to a common brown-olive color.   Yet some of the Chemung shales are found, when freshly broken or dug into, to be green, olive, dark and even deep black; and the broken sandstones are of a light grey.   A general tinge of green or olive pervades the whole formation.*

The *upper part* of the formation is characterized by a general tendency to conglomerate, or gravel.   In a few localities the mass becomes a well characterized pudding stone.   In these gravel beds the same animals continued to live and breed, the remains of which we find in the shales and sandstones lower down.

"This conglomerate (continues Mr. Hall) nowhere attains sufficient thickness or importance to merit a distinct description; but in hasty observations it may, sometimes, lead to erroneous inferences, since it resembles in many respects the distinct and well-defined *conglomerate which rests upon this group* in

---

* In the old survey we always spoke of the olive shales of VIII, and the green sandstones of VIII, meaning Chemung.

the western part of the State, *but which is totally distinct from the same.*"*

Many of the shaly sandstones and shales of the Chemung group are highly micaceous;† and towards the top of the group the shales are coarse, fissile and reddish, with much mica in glimmering scales, with a slight change in the character of the prevailing organic forms.

The Chemung Formation forms the high hill country bordering the New York and Pennsylvania State line on the north and the south, at an average general elevation of 2,000′ above sea level, or about 1,500′ above Lake Erie.‡

North and south lying valleys cross this highland from the one State into the other, which are excavated to various depths of from 500′ to 900′ below the general upland surface, and have usually steep or precipitous side slopes. The upland country consists therefore of a series of parallel north and south lying, broad, flattopped but undulating ridges, topped or faced above by the Upper Sandstones of the Chemung Formation, and set with low knobs composed of isolated patches of Chemung conglomerate. The valley beds and foot slopes exhibit nearly horizontal, but gently southward dipping, middle and lower Chemung strata.§

---

* This is the earliest plain statement made by any geologist touching the Oil-Sands and Mountain-Sands of Pennsylvania. It was published in 1843. Twenty years afterwards the oil borings furnished elements for a thorough discussion of the problem which Mr. Hall thus early hinted at, but had no adequate means for fairly stating, much less solving. It must be carefully borne in mind that the description of Chemung rocks in the Report of 1843, was got chiefly from the middle region of southern New York, along the Chemung river. In the sentence just quoted, where Mr. Hall contrasts the Chemung conglomerate ("this conglomerate" of the text) which has no special importance, with another conglomerate *in the Western* part of the State, he means the Ellicottville, Olean, Chautauqua conglomerate, supposed at that time to be the Great Coal Measure Conglomerate No. XII.

† The finer fragments of silvery mica floated, as we have seen, as far as the extreme knife-edge limit of the formation, west of Cleveland, in Ohio. They prove that the Chemung rocks came from rivers flowing through the ancient continent of Canada, New England and the Atlantic region, made up, to a large extent, of mica-slate formations of extreme antiquity.

‡ 573.08 above tide. (Gardner's last determination.)

§ The patches of conglomerate which occur as top coverings of the highest hills are fragments of a deposit extending into and under Western Pennsylvania. The preservation of these patches has been due to one or both of two causes. Either an original exceptional thickness, massiveness or coarseness of the conglomerate made it at these points able to endure longer the

The following levels taken from Plate 12 of Prof. Hall's Final Report of 1843 will explain the description just given. These levels follow an irregular east and west line rudely parallel with the State Line and to the north of it, extending for 150 miles from Portland Harbor on Lake Erie to Ithaca at the head of Lake Cayuga :—

Section from West to East.

Levels above tide.   1,000' 1,200' 1,400' 1,600' 1,800' 2,000'

Lake Erie   573'
  Summit - - - - 1,364'
Chautauqua Lake - - 1,299'
  Summit - - - - - - - - 1,974'
Conewango creek - - 1,258'
  Summit - - - - - - - - - 2,143'
Elliottville - - - - - 1,522'
  Summit - - - - - - - - - 2,152'
Oil Creek, New York - - - 1,452'
  Summit - - - - - 1,495'
Angelica on Genesee river - 1,436'
  Summit - - - - - - - - 2,070'
Arkport - - - 1,202'
  Summit - - - - - - - 1,848'
Bath - - - 1,098'
  Summit - - - - - 1,587'
Mud lake - - 1,119'
  Summit - - - - - - 1,652'
Jefferson   464'
  Summit - - - 1,265'
Ithaca   418'

To the above may be added from the same source :
Crooked Lake 726'
Casadaga Lake - - - - - - 1,699'
Lime Lake - - - - - - 1,631'
Allegheny River at the South end of Hall's section :
Plate XI.   Report of 1843 - 1,289'.

---

slow erosion of atmospheric agents ; or else the patches lie in long wide gentle synclinal troughs, scarcely discernible to the general observer, but subject to instrumental investigation, which they must soon receive. It is probable that the conglomerate deposit in question varied locally in coarseness, as we now know that all the Mountain Sands and Oil Sands do, of which this in fact is the most important one.

The *lower* Chemung beds are especially well exposed a little south of the heads of Seneca and Cayuga lakes; and the *upper* Chemung beds, along the river hills further to south.

The whole formation varies in its character geographically. Towards the east the strata are darker and more subdivided into shales and sands, so that distinct and well defined beds are not so common. The shales are of a dark-olive color, and black shales appear only along the State Line. But further west, on the Genesee river, the shales are often in thick beds, bright green, without interbedded layers of sand; and the sandstones are also of a lighter color, and less intermixed with shaly matter.

Six divisions are made by Hall in his Genesee section of the Chemung:

6. *Falsebedded Sandstone and Conglomerate.*
5. *Old Red Sandstone.*
4. *Grey and Olive Shales and Shaly Sandstone.*
3. *Green Shale with Grey Sandstone.*
2. *Black slaty Shale with Septaria (concretions.)*
1. *Olive Shale Sandstone.*

Under this Portage Sandstones.

No. 3. *Green shale* is the principal member of the group everywhere, and becomes more and more predominant going west. Its sandstone interpolations almost altogether thin away; although in Cattaraugus county there are some thick masses of greenish gray sandstone, very durable, and readily quarried into large blocks. [These possibly represent the Oil-Sands of Pennsylvania, but without their streaks of gravel.] The dark olive sandstones of Steuben county, in Middle New York, are not to be recognized in Chautauqua county, at the south-west corner of New York, except as thin layers of brownish sandy shale, containing the same fossils.

The red sandy micaceous shales of the Chemung do not cross the Genesee river westward.

There is an evident thinning of the whole Chemung Formation westward. There is also a constant decrease in the number of its fossils; many common ones in middle New York not being found towards lake Erie; and scarcely any new forms appearing to take their place.

But there is an immense preponderance in the number of fossils, (both of species and individuals) in the Chemung Formation over the number of fossils, (species and individuals) in the preceding Portage age, when the ocean was apparently deeper. The depth of water had evidently become so much diminished by the deposit of the Upper Portage Sands that shell-fish could flourish in the Chemung sea.

A very few fragments of floated land plants have been detected in the Chemung rocks of Middle New York; but none further towards the south-west. The land from which these relics of land vegetation floated out must have been in Eastern New York or New England.*

Many of the thin sandy laminæ are often almost completely covered with small fragments of coaly matter, apparently some land vegetation, broken up by rivers, and floated far out as a film on the sea surface, to be deposited at last, with the sand and mud, at the bottom of the sea; as the water surface of modern bays and lakes are sometimes seen covered for great distances with a thin scum of ground-up decayed wood. The earth was evidently getting ready for the Coal Era, preparing as it were for its great Centennial Exhibition, in honor of what it had a hundred ages previously but partially accomplished in the Coal Formation of the Hamilton age in Middle Pennsylvania.

Mr. Hall gives the following as a specimen of a thousand such sections to be made along any of the hill slopes of the north and south valleys of southern New York:

---

* Note what Prof. Rogers says of these in page 811, F. R., where he describes the "fossilized stems of what seem to be coniferous trees," as they appear in the Appalachian Coal Measures. He says they "almost invariably lean at a low angle in the thick sandstone strata which enclose them; they seldom occur in the shales, but are nearly restricted to those coarser rocks which appear to have been deposited by rapidly-flowing currents. It is not a little curious," he adds, "that nearly all of these trunks dip towards an opposite quarter from that towards which the false bedding or oblique lamination of the surrounding sandstone itself declines. This fact noticed by me not merely in Pennsylvania and on the Ohio River above Gallipolis, but also in the instance of the celebrated Craigleith tree near Edinburg, and in the great tree of the Granton quarry, near the Firth of Forth, goes far, I think, to demonstrate that all these leaning stems have been imbedded in the manner of *snags* in actively-moving currents, their butt-ends settling first, and their lighter tops pressed forward by the waters, which at the same time piled the sand obliquely about them with a forward dip."

Section of Chemung Narrows.

16. Shale, olive, fissile, with *Aviculæ,*   -   -   - 15' 0''
15. Shale, compact with *Cyathophylli,* and other corals, 0' 6''
14. Shale, compact, with thin courses of sandstone
     separated by seams of shale, -   -   -   - 13' 0''
13. Sandstone, greenish gray, with seams of shale, - 10' 0''
12. Sandstone, greenish gray, weather-edges stained
     with oxides of magnanese and iron,   -   - 7' 0''
11. Shale and sandstone, with *Aviculæ, Atrypæ,* &c.,   5' 0''
10. Shale, soft, greenish olive, -   -   -   -   - 3' 0''
9. Shale, compact, sandy, with *fossils,*   -   -   - 2' 0''
8. Shales in three distinct courses, 2,' 4' and 6',   - 12' 0''
7. Coral animal remains (*Corallines,*)   -   -   - 0' 2''
6. Shale, olive, with abundance of *fossils,*   -   - 3' 0''
5. Sandstone, compact, shaly,   -   -   -   - 2' 6''
4. Shale and thin sands ; abundance of *fossils,*   - 6' 0''
3. Sandstone, concretionary, -   -   -   -   - 3' 0''
2. Shale, with thin layers of sandstone, -   -   - 8' 0''
1. Down to river-bed concealed,   -   -   -   - 14' 0''

The above section was made at the Chemung Narrows, in a cliff showing 90' of outcropping rocks. The principal fossils to be found so abundantly here, were named by him: *Avicula pecteniformis, Strophomena membranacea, Strophomena interstrialis, Orthis interlineata, Delthyris prolata, Atrypa aspera.*

In another section at Corning, of rocks very similar in character, scarcely any of the above fossils can be found, but instead of them *Cypricardia, Avicula spinigera, Dethyris* —, *Orthis* —, and plenty of *Orthis unguiculus, Orbicula, Loxonema, Tentaculites,* &c. Every inch of both sections can be measured, and there can be no mistake about the totally different aspect of the animals inhabiting the sea at the two localities. Similar sections at other localities, showing some variation in the character of the deposits, show also a considerable difference in the species of fossil shells which inhabited the Chemung sea in different parts of its bed.*

Cross lamination, or oblique, or current bedding,—ripple

---

* The Chemung fossils will demand attention in a future chapter. See Hall pp. 261ff.

marks,—concretions, &c., are abundant in the Chemung, as in the Portage.

The mineral contents of the Chemung formation are, with one exception, of little interest. No important beds of coal, iron or limestone; no veins of copper or lead occur in it. Petroleum takes the place of all these, and makes the Chemung Oil-Sand rocks famous over the world.

Iron sulphide (pyrites) must be extensively distributed throughout the shales, judging by the amounts of alum and copperas (sulphates of alumina and iron) set free. Many shells are found, all the lime of which has been dissolved away as gypsum, and its place filled with a cast of iron pyrites.

In like manner iron-carbonate (spathic iron ore) fills the casts of crinoid stems and other fossils in the upper rocks; rarely those lower down.

Manganese oxide, or black wad stains the rocks, and collects at the foot of every cascade.

The best localities for studying the Chemung rocks in middle New York are at Ithaca, on the inclined-plane of the railroad; at Cascadilla Falls; and on Falls creek; for the Lower Chemung. On Cayuta creek; at Chemung Narrows; near Elmira; near Bath; and at Corning, along the Blossburg railway cutting, for Middle and Upper Chemung.

Along the Genesee river on Caneadea creek (green shale); at Rockville on Black creek (rich fossil green shale); at Hull's mills near Angelica; Hobbieville; Phillipsburg (same fossils as at Rockville); on Vandemark's creek (higher strata loaded with *Delthyris,* good exposures, fine specimens); at Wellsville (Old Red IX comes in, capped by gray sand conglomerate X.)

In Cattaraugus county good opportunities for examining the Chemung rocks occur at Bailey's in Leon; two or three near Cadiz; in the southern part of Great Valley; and in the ravines on the south side of the Allegheny river (but few fossils.)

In Chautauqua county, the deep ravine of Chautauqua creek is the best place, and affords many fossils. At Dexterville, the outlet of Lake Chautauqua, are rocks loaded with fossils.

The Chemung formation in middle New York can scarcely be less than 1,500′ thick, estimated by a dip along a N. and S. line 30 to 40 miles long, with hills of 800′ at the State line.

On the Genesee river, the Chemung formation must be above 1,500', calculated by Professor Hall thus:—

		Above tide.
Top rocks of the Portage formation,	- - less than	1,200'
Lowest passes in the hills further south,	- -	1,500' to 2,000'
Highest hills near the State Line,	- - -	nearly 2,500'
Difference of top of Portage and top of Chemung,	-	1,300'
Add for an undulating dip southward, -	- at least	200'
Total least possible thickness of Chemung,	- - -	1,500'

On Lake Erie the Chemung Formation has thinned visibly and considerably, but has not been carefully measured. The difficulty is to find a top layer to measure down from. There are few localties where the junction of the Chemung VIII, with the Old Red Sandstone IX, can be clearly seen; and the three formations IX, X and XI, which intervene between the Chemung VIII and the Coal Conglomerate XII as mountain masses in Middle Pennsylvania, and which are still of very considerable size in Tioga and Potter counties, become so thin and so changed in character, after passing the Genesee River and approaching the Lake Erie region, that it has required all Mr. Carll's skill and care to unravel the difficulty, and lay a new basis for settling the question of the thickness of the whole Chemung Formation in South-Western New York.

CHAPTER XII.

The Formations above the Chemung in New York. IX, X, XI.

The only formation which, in 1843, Prof. Hall recognized as intervening between the top of the Chemung and the Coal Measures, was "a thin band of the Old Red sandstone, a highly ferruginous stratum, containing few fossils."[*] "In other places[†] the

---

[*] Report, page 276. See Section on his page 253. See his Plate XI, section across Allegheny county, New York, showing the top conglomerate of the Chemung; and Plate X section across Steuben county, New York, showing "Old Red" resting on Chemung.

[†] See Hall's Plates XI and XII, Report of 1843, where the Conglomerate No. XII rests on Chemung No. VIII.

Conglomerate of the Coal Formation rests directly upon the Chemung group."

" Where the best opportunities for examination exist the change from the Chemung to the Old Red sandstone is abrupt in character.   The greenish and olive shales and sandstones charged with *Strophomena, Delthyris* and *Atrypa,* are succeeded by a red sandstone containing *none of the organic remains of the lower rocks.* * * *   Few fossils are known; but the rock everywhere contains the remains of *fishes* which are often preserved in a very perfect manner. * * *

"The Old Red Sandstone is scarcely known west of the Genesee River." * * * " In many places the conglomerate [by this is meant the Great Coal Measure Conglomerate No. XII,] rests upon the Chemung group, without the intervention of any other rock."

The writer can fully sympathise with Prof. Hall's embarrassments in settling the relations of the Chemung to the Coal Measures for his final report of 1843 ; for he himself encountered the same difficulties and made the same erroneous statements in reporting to Prof. Rogers on the northern tier of counties in Pennsylvania in 1841.

The deceit which deceived us all, and has deceived everybody, ever since, is the pretense of the Second Mountain-Sand of Venango county, the Garland conglomerate and Great Bend conglomerate of Warren county, Pennsylvania, the "Rock City" conglomerate west of Olean, at Ellicottville and elsewhere in Cattaraugus county, New York, the " Rock City" conglomerate of Chautauqua county west of Lake Chautauqua (for by all these names is it known) to be the Great Conglomerate, Coal Measure Conglomerate No. XII, Rogers' Seral Conglomerate, the Millstone Grit of the English and the Farewell Rock of miners sinking deep shafts to and through the Lower Coal measures.

Whereas, the northern edge of the Great Conglomerate No. XII seems nowhere to reach the New York State Line, even in outlying patches ; but thins away northward, and has been washed from off the face of the counties in both States, both north and south of the Line.

A rock stratum, however, lying 200 feet beneath it, thickens and becomes a great conglomerate in all that country, and plays the very part which No. XII ought to have played; preserving high ridges and sharp knobs, forming precipices, rock cities, &c., just as No. XII does wherever it exists. No wonder then that the one was taken for the other, the lower false conglomerate for the upper true Conglomerate, the Vespertine White Sandstone No. X for the Seral Conglomerate No. XII.

The following scheme will show the old problem and its recent probable solution:

In Ohio.	In N. W. Pennsylvania.	In New York.	In Middle Pennsylvania. Rogers.	
Conglomerate.			Seral Cong.,	XII.
Cuyahoga shale.*			Umbral,	XI.
Berea Grit.	2d Mtn. S.	Conglomerate.	Vespertine,	X.
Bedford shales.		Old R. S.S.(fish).	Ponent,	IX.
Cleveland shales.	Oil Sands.	Chemung.	Vergent,	VIII.
Erie shales. †		Portage.	Vergent,	VIII.
Huron shales.		Hamilton.	Cadent,	VIII.
Corniferous.		U. Helderberg.	Postmedidial,	VIII

CHAPTER XIII.

The Old Red Sandstone of New York, IX.

In Middle Pennsylvania the Old Red Sandstone of England is well represented by a very thick formation (No. IX, Ponent of Rogers) constituting ranges of mountains, such as Terrace Mountain and Sideling Hill of Blair and Huntingdon counties; the Cove Mountain, Peters, Mahontongo, Nescopec, Shickshinny and North Mountains, on the west side of the

---

* I have not used the general term Waverly Sandstone Formation of the Ohio Reports, because of the controversies to which it has given rise (See among other writings, 23d Ann. Rep. of the Regents Univ. New York, Albany, 1873, J. Hall, pages 7 to 10) and because its subdivisions correspond to our Pennsylvania formations. Dr. Newberry, in his Report of Progress of the Ohio Survey for 1870, page 59, divides it into three members, upper, middle and lower; the middle Waverly being a conglomerate. This is the Berea Grit and the New York "Conglomerate;" our Venango Second Mountain Sand. But in the Report of 1873, the Ohio geologists divide the Waverly Group into four formations, Cuyahoga, Berea, Bedford and Cleveland.

† This should stand opposite to both Chemung and Portage, in the other column as the upper Erie contains Chemung fossils.

7—1.

anthracite coal fields ; and the Second Mountain, which borders the anthracite field on the south in an almost straight line from Harrisburg to Mauch Chunk, and is continued eastward as the Pocono Mountain to the Delaware River, and as the Catskill Mountain to the Hudson River, a distance of two hundred miles. In the same manner the North Mountain of the Harvey's Lake and North Branch Susquehanna country forms a continuous Old Red Sandstone wall, under the name of the Back Bone Allegheny Mountain, past Muncy, Williamsport, Lock Haven and Altoona to the Maryland line, where it becomes the Great Savage Mountain and encircles the Cumberland coal basin. Hence it can be traced through Virginia into Tennessee.

The Old Red, or Red Catskill, Formation No. IX, is between 1,000 and 2,000 feet thick where it outcrops south of the anthracite coal fields and on the south flank of the Catskill Mountains in New York. But its thickness is reduced to 600′ and 400′ in the northern counties of Pennsylvania, where it forms the Elk, Towanda, Blossburg and Cowanesque Mountains and the north flank of the Catskill Mountains in New York. As it is buried under the bituminous coal fields of Clearfield, Jefferson and Elk counties its thickness there is unknown; but by tracing its northern outcrop westward through Potter, M'Kean and Warren in Pennsylvania, and through the southern tier of counties of New York to Cattaraugus county on the Genesee, and Chautauqua county on Lake Erie, it is plainly to be seen that the formation as a whole thins down to less than 100 feet along lines directed north-westward towards Lake Erie. Its materials must have been derived from the highlands of New Jersey, the South Mountains of Eastern and Southern Pennsylvania, and the Blue Ridge of Virginia.

The geologists of Ohio do not recognize its existence in that State. But it is represented by the red Bedford shale of the Waverly group, underlying the Berea grit, (No. X, Vespertine, White Upper Catskill.)

This general statement will suffice, at present, to make plain Prof. Hall's description of the Old Red Sandstone of the Fourth (Western) District of the New York survey, published in 1843, (pages 278 and following,) an abstract of which will now be given.

The Old Red Sandstone (he says) where fully* developed, (i. e. in eastern New York,) consists of various strata of sandstone, shale and shaly sandstone, conglomerates and impure limestones.

The prevailing color of the sandy parts is brick-red, though often lighter, and sometimes of a deeper color, from a larger proportion of iron; while the coarser parts are often gray, and the shales are green.† Beds of green shaly sandstone are interstratified with the red friable sandstone, and these are succeeded by a compact kind of conglomerate rock.

But in the Fourth District, or south-western counties of New York, this formation is of no great thickness or extent.‡

Along the Genesee River, and in some of the higher hills south of the Canisteo in Steuben county, this formation consists of a thin mass of calcareous sandstone, highly charged with iron, and containing the remains of fishes.§

West of the Genesee Prof. Hall detected this rock only in loose masses upon the surface, though it probably occurred at some of the elevated points, escaping observation by being covered with forests.

Further west, where good opportunities for observation existed, it was not found, and " it thus becomes evident, that in this direction the rock disappears not far from the Genesee in Allegheny county," a fact which Prof. Hall ascribed to "a diminution in the transporting power of the oceanic currents which had previously carried forward similar materials over the broad extent before described" that is, "nearly or quite as far as the Mississippi River."

" Still further west sandstones and thin limestones overlie the Chemung (No. VIII) wholly unlike the red and green sand-

---

* It is not as fully developed in eastern New York as in Middle Pennsylvania, where it has been well described in Prof. Rogers' Final Report of 1858, and will be carefully described in Mr. Dewees' Report of Progress for 1874.

† A continuation of the deposit of the green and olive shales of the Chemung, No. VIII.

‡ What is added on p. 278, Hall's Report, has been corrected by Prof. Hall's recent publications, and is therefore omitted here.

§ Prof. Hall says that while the lithological character of the Catskill corresponds to that of the English Old Red, it also contains at least some of the same animal fossil forms as in England. I believe that I was the first to find an American Old Red *Holoptychus* in 1840, in Tioga county, Pennsylvania. But the Old Red fossils were fully recognized in the New York Geological report of the work of 1840 in that State.

stones and sandy shales of No. IX (Catskill) in New York, and they contain at the same time a different assemblage of organic remains. Other changes too have supervened at the west, of which we have no evidence in New York; and the most striking is the occurrence of an important mass of limestone [XI] below the conglomerate [XII] which is the great supporting rock of the Carboniferous system."

This quotation is given to bring out more clearly our present knowledge of the facts, which draws an entirely different picture of the situation. The last sentence is the key to the old error. What was called Conglomerate (XII) in south-western New York turns out to be Upper White Catskill (Vespertine; Second Mountain S. S.; X;) and the limestone of the west lies *above* it, not below ; and *therefore* does not appear in New York. The Old Red, instead of thinning out and disappearing, certainly continues on through north-western Pennsylvania into eastern Ohio, and carries its fishes and other organic remains along with it.

Much however remains to be done before our knowledge of this important formation can be satisfactorily arranged. The great obstacle in our way still, is the presence of *red beds in the top of VIII*, called by me, in my report of 1840, the Red Mansfield Beds of Tioga county. It was impossible at that time to trace these beds through the wilderness of Potter and M'Kean to Warren county, Pa., or through the New York counties to the Genesee river, and keep them separate from the true Catskill red beds so closely overlying them. So that no one even now knows which is which in M'Kean and Warren counties; and it is an open question still, whether Mr. Hall's description of the Old Red on page 280 of his Report of 1843 be not of the Mansfield red beds of VIII (Chemung) instead of the outcrop of the red rocks of IX (Catskill.) The description which follows corresponds very closely with the character of the Mansfield beds, and does not correspond to the character of the Red Catskill anywhere in Pennsylvania.

"In the Fourth District, the thinning margin of the Old Red Sandstone, from the large proportion of ferruginous matter it contains has usually more the appearance of an iron ore than a sandstone. It appears to be a compound of sand, clay and calcareous matter, with a large proportion of hydrate of iron, and contains, in abundance, small fragments of bones and scales of fishes. These are generally too small and too much worn to be recognized, except in the general similarity with better characterized specimens of scales and bones of *Holoptychus* from other localities. In some places the whole mass is an iron ore of tolerable quality, containing probably 20 or 30 per cent of that metal. In such cases however it is very thin and *I have been unable to find any rock about it in connection.**

To settle this question, if possible, Mr. Andrew Sherwood has been appointed to survey the district of the north-east counties, beginning with Tioga, where the Mansfield beds are best developed. Mr. Sherwood's residence is there, and he has had some years of geological experience in this very matter, having been assistant geologist both in New York to Prof. Hall, and in Ohio to Prof. Newberry. He and his brother are making a special survey of the outcrops of the Catskill and Chemung, laying them down in colors on the county maps, and marking the occurrence of the ore beds, of which there are several. (July, 1875.)

The principal localities marked by Prof. Hall as Old Red are near Wellsville, on the Genesee River, and at another point near Spring Mills in the south-eastern part of Allegheny county. It likewise appears on several hill tops in Steuben county between the Canisteo and the State line.

The green shales and shaly sandstones of the top of the Chemung, VIII, are charged with shells of *Delthyris, Strophomena* and *Atrypa.*

"Nothing like these occur in the rocks next above (Hall, 1843;) the shells being quite distinct. Fish bones and scales are characteristics of the Old Red, *meeting the eye in all localities* from their strong contrast with the ground in which they are imbedded, *being usually white or bluish* in the brownish red rock. These scales and bones are often in minute fragments, and so permeated by iron as to offer no contrast in color; but in other places they are perfect, and from 1″ to 1½″ in diameter, appearing like patches of extraneous matter."

---

* I have underscored these words, because by the rules of topographical geology, they would be in evidence against the deposit being Catskill and in favor of the deposit being Mansfield.

" By far the most common of these are the scales of *Holop-tychus nobilissimus* well known in Great Britain, the enamel marked by large wavy furrows and ridges, rubbed off as it were on the part of the scale overlapped by the next scale. The casts of the enamel are often preserved when the enamel itself has been weathered away. Teeth like those of *megalichthys* are often found. A jaw (set with teeth) 7 inches long is figured on page 282 (Hall, 1843.) To a fin in a lower bed, Prof. Hall gave the name *Sauripteris.*

This is all respecting the Old Red to be found in the New York report of 1843 of the Western district. Copious materials have been pouring into Prof. Hall's cabinet of late years, and the study of the fossils already collected in the Northern counties of Pennsylvania, and a comparison of them with the very splendid collections of Dr. Newberry from Ohio, will in a short time enable us to speak clearly and definitely respecting what has hitherto been so darkly confused.

CHAPTER XIV.

The Supposed Conglomerate in New York.

" The Old Red sandstone in the Fourth District (western part) of New York is succeeded by a coarse siliceous conglomerate and a gray diagonally laminated sandstone, the former generally prevailing. The conglomerate consists of a mixture of coarse sand and white quartz pebbles, varying from the size of a pin's head to the diameter of two inches. They are generally oblong, or a flattened egg shape. Some of these are of a rose tint when broken, but white upon the exposed surface. Pebbles of other kinds are very rare in the mass, though red and dark colored jasper are sometimes found.

" This rock in the Fourth District occurs in outliers of limited extent, capping the summits of the high hills toward the southern margin of the State. It is represented on the [old geological] map [of New York] by small dark spots, and its relative position is seen by an inspection of the sections crossing the counties, (Plates 10, 11 and 12.)*

---

* Hall's Report of 1843, Plate X, only shows the Old Red capping the two highlands of Troupsburg, south of the Canisteo river in Steuben county, and

"From its position it has been undermined; and separating into huge blocks, by vertical joints, which are often many feet apart, the places have received the name of *Ruined City, Rock City*, &c.* In many situations it can hardly be considered as being in place, the wearing away of the rocks beneath having allowed the mass to fall down, so as to occupy the side of a hill instead of the summit. It is often much broken; and scattered fragments extend on all sides of the principal outliers to considerable distances. In many instances I have detected huge fragments of this rock nearly as far north as the northern limit of the southern range of counties, lying on the hill sides, and sometimes in the valleys. At first I was disposed to consider these as transported from the south, knowing no rock in place of the kind so far north upon its surface.†

"Subsequent investigations, however, have convinced me that these fragments‡ are the remains of the rock itself, which once extended continuously much farther north than its most northern outliers at the present time. §

---

the highland next south of it. Plate XI shows the conglomerate in Allegheny county overlying Old Red on the hills south of Wellsville; but the hill north of Wellsville and south of the Genesee river has only Old Red. In Cattaraugus county the section shows the conglomerate (without any Old Red) capping the hill north of Ellicottville; the hill south of Ellicottville and north of the Allegheny river; and the hill south of the Allegheny river. Plate XII shows the conglomerate in Chautauqua county on the hill tops on both sides of Chautauqua lake, and on that south-west of the valley in which Panama lies.

* A sketch of the most famous of these rock cities, 6 miles south-west of Olean, by E. N. Horsford, is given at the head of the chapter; but it must be accepted as a *diagram* rather than as a landscape portrait of the place. It shows the structure, however, very well.

† Suggesting the agent of erosion. But care must be taken to distinguish between that frost and water action through all the ages since coal, which did ninety-nine-hundredths of all the work, and that glacial action so much talked of now which merely scratched and polished the last remaining eroded surface. These striæ and scratches occur on the highest hill tops in Cattaraugus county, nearly 2,000 feet above similar markings on the rocks along the shores of Lake Ontario. Agassiz was the first to teach us how the northern cope of ice flowed over the continent in a sheet several thousand feet thick, overtopping our highest mountain crests.

‡ Very striking but rather artificial looking pictures are given on pp. 286, 287 and 288, report of 1843, of isolated masses of this conglomerate to show the vertical cleavage and the diagonal or current-bedding.

§ Professor Hall adds, after giving some more facts, and alluding to the plausibility of the iceberg theory: "Although this would show a former extension of the conglomerate to points twenty miles north of where it is now

"The diagonally laminated structure is often beautifully pre-sented in the broken cliffs and large fragments of this rock, and this is one of its most obvious characters."*

Now if there is one feature of the White or Upper Catskill, No. X, more characteristic of it than any other, throughout Pennsylvania, in all the mountains composed of the great rock, especially along the whole 200 miles of face of the Allegheny mountain, and in the gorges of Laurel Hill and Chestnut Ridge in

known to be in place, the inference seems unavoidable and well substantiated."

Although these phenomena of outcrop erosion strike the imagination very forcibly in the case of marine sandrocks, producing rock cities, &c., they differ in no other particular from less remarkable but strictly similar outlying patches of any other kind of rock formation. There is every reason to believe that Lake Erie was eroded out of all the measures from the Coal down to the Trenton, just as Nittany Valley or Kishacoquillas Valley in Central Pennsylvania has been. No one doubts that the Niagara rocks once stretched over into Canada. No one doubts that the Chemung formation of the south shore of Lake Huron once formed an unbroken layer across both lakes and the intermediate peninsula of Upper Canada. If then the Second Mountain Sand of Venango county, Pennsylvania, can be followed two hundred miles southwestward into Ohio as Berea Grit or Waverly Sandstone, and three hundred miles eastward to the White Sand Summits of the Catskill Mountains, and fifty miles northward to its last fragments left on the hill tops, half way to Lake Ontario, there is small reason to doubt that it once covered the area occupied now by the lake; and perhaps bore the Coal Measures of Pennsylvania on its back.

Professor Hall was led by a similar natural geographical train of reasoning *towards* the same conclusion. He says: "The prominent outliers of this rock upon the high and distant hills furnish good landmarks for showing the extensive devastation which this part of the country has suffered. From its great extent northward, (i. e. the conglomerate,) it appears very probable that originally some of the lower beds of the Pennsylvania coal-fields extended into New York, and that being of more destructible materials than the conglomerate, they have been entirely swept off. This may appear more reasonable when it is stated, that about six miles south of the State line, on a high ridge of land between the Allegheny River and Conewango Creek, *a bed of coal lies upon the Conglomerate*, which latter extends thence northward into New York, its broken outliers appearing for ten or fifteen miles north of the State line." This is the same coal seen near Warren, Pa., and all through south Crawford county, and belongs to the Sharon Coal series, *under* the true Conglomerate No. XII. The same coal *underlies* the great Conglomerate No. XII throughout Tioga, Bradford and Lycoming counties, at Towanda, Ralston, &c.

* Fig. 133, in Prof. Hall's report, of a quarry at Burned Hill, south part of Allegheny county, shows the current bedding in the alternate great layers of the rock. The intermediate layers are perfectly regular and horizontal. A still more striking view of this current-bedding on a grand scale is shown in fig. 134, in a block of the conglomerate at Rock City south of Ellicottville.

Somerset and Fayette counties, it is precisely this feature of current-bedding. Hence in the Old Survey of 1835–41 we were accustomed to speak of this formation as the " false-bedded Sandstone mass of No. X."

There are numerous instances of false-bedding in the Conglomerate No. XII, and in the Mahoning and other Sandstones of Coal Measures above it. But they are sporadic and rudely irregular; whereas the current-bedding of No. X is universal and beautifully distinct. One of the finest exhibitions of it was (and perhaps still is) in a block of **X**, weighing many tons, which had fallen from the cliffs and rolled down upon the bottom land of the Conemaugh river, near the foot of the railroad embankment at the west end of the high Viaduct above Johnstown. If cut square it would make a beautiful pedestal to a statue, without any further ornamentation than that bestowed upon it by nature in the lines thus marked upon its surfaces.

"In some places this rock is traversed by seams of iron ore, which often stand out from the surface of the blocks, having resisted the influence of the weather which destroys the mass. These seams are seldom in right lines; at other times undulating; and rarely if ever, in the lines of deposition. They are evidently segregations from the mass, and the irregular and undulating direction is due to accretionary force.* In some instances nodules of hydrate of iron have been found; at other times, the folded seams or laminæ enclose a mass of sand, which has apparently resisted the formation of a solid concretion of iron ore."

The localities in western New York, where this No. X White Catskill conglomerate can be studied, are given by Prof. Hall, on page 289.

Near Wellsville on the Genesee.

In the town of Scio, 4 or 5 miles further west, where millstones are made of it, (compare the quarries of Berea Grit in Ohio.)

In the town of Little Genesee; rock coarse like the last.

Six miles south of Olean, nearly on the State line; several acres; north outcrop much broken; huge masses scattered for a mile or two around. Some of the undisturbed blocks 70'×40'

---

* Figs. 135, 136, 137 curiously illustrate this phenomenon. The last represents Mr. Hall's observations at Cuyahoga Falls in Ohio, which he describes in Trans. Assoc'n. Amer. Geol. and Naturalists, Vol. I, 1843.

on a base, with chasms like streets between; once a great bear and wolf den.

At Rock city, 7 miles south of Ellicottville, in Cattaraugus county, on a hill top; blocks widely scattered along the brow; main body of about 50 acres nearly undisturbed, but chasmed by the weathering of the joints; as if the breakers of a sea once washed in among them.*

Near Judge White's, a few miles south-east of Ellicottsville; two hill tops; separated by an east and west valley; scattered blocks.

Between Napoli and Little Valley; blocks.

Between Little Valley and Great Valley; blocks.

Several hill-tops south of the Allegheny River; scattered blocks.

Two miles west of Ashville in Chautauqua county; quarry of gritstone.

William's quarry, four miles north of Panama; much rock has been taken from here.

One mile north-west of Williams' quarry; another quarry associated with a few inches of conglomerate rock.

At Panama; masses along the eastern slope of the hill, and on both sides of the stream. Rests on soft green shale, the destruction of which has allowed the mass to slide down from its original place. Blocks 70' long, 40' wide, 70' high. Rock nearly continuous, only separated by fissures.

North of Chautauqua Lake, on Mr. Young's land; loose blocks.

Between Ellington Centre and Cassudaga Creek, on Mr. Barnard's and Mr. Preston's lands; loose blocks.

Three miles south of Ellory Centre, on Mr. Strong's land; loose blocks.

---

* If the south dip of 20' to 40' to the mile now existing were restored to the horizontal, without elevating western Pennsylvania, this 2,000' hill would be brought down to tide level. Some such restoration of the original horizontal is necessary to account for the evident former drainage of North Pennsylvania waters into Lake Ontario; the excavation of the smaller New York Lakes; and the excavation of Lakes Ontario and Erie themselves. But this subject will receive separate treatment elsewhere. A stiff sketch of this locality is given on page 290, fig. 138, and it represents the rock as over 30 feet thick, divisible into five distinct layers, the lowest being beautifully current-bedded.

Four miles north-west of Panama, at the foot of a hill on Mr. Field's land, mostly small fragments of conglomerate and sandstone cover the ground to some depth, as if they had rolled down from a high eminence.

In Clymer, several miles west of the last, is a grindstone quarry.

About three miles south-east of Panama, on the east side of the Valley of the Little Broken Straw; conglomerate on Mr. Lloyd's land.

On lot 13 still further east, on Mr. Vosburgh's land; two or three acres of sandstone, outcropping on the north and east sides of the hill; covered with a layer of fine beach sand; rock surface worn and smooth.

In the south-east corner of Chautauqua county, two hills one on each side of Case Run, are topped by this formation consisting mainly of sandstone, with a little conglomerate, huge blocks, 30 feet high, scattered over the tops and sides of the hills. The rock faces the divide between Conewango Creek and Allegheny River from this place nearly all the way southward to their confluence, and a coal bed comes in over it about six miles south of the State Line in Pennsylvania.*

The thickness of the conglomerate is generally between 25′ and 35′ rising to over 60′ at Panama.†

" Fossils are extremely rare in this rock, having been seen in one locality only, four miles north of Panama; and in this, sandstone greatly predominates over conglomerate. Fig. 139, p. 291, shows what these fossils are, viz: *Euomphalus depressus*, (a spiral shell;) *Cypricardia? rhombea;* and *Cypricardia contracta*, (beaked shells.) All of them small.

It only remains to add an important reference which Prof. Hall makes, in a foot note to page 286 of his Final Report of 1843, to the conglomerate in the Chemung rocks lying at a much lower geological level than the one described above.

" There is another conglomerate in Chautauqua county, and in some places in Allegheny county, which was briefly noticed under the Chemung group. This, however, is a thin mass, and

---

* These data are obtained from the annual report of Mr. Hall for 1841.

† Mr. Hall states its thickness in Pennsylvania at 150′; but this is due to its being mistaken for the Great Conglomerate No. XII.

wherever it has been found in place, is associated with fine-grained compact sandstones, and frequently contains the fossils of the Chemung group. In the northern part of Chautauqua county, I found some loose masses of this conglomerate containing fossils known to belong to the Chemung group, and by this they were chiefly identified. The aspect of the rock is also somewhat different; the pebbles smaller, more round, and not of the same white quartz which occurs in the higher rock."

It is evident that we have here the conglomerate which Mr. Sherwood has recently traced along the *top* outcrop of No. VIII (Chemung) through Tioga county and further west, *underlying* the Old Red (Catskill,) and closely connected with the Mansfield red beds.

It is undoubtedly one of the thin conglomerates outcropping at Warren, Pa. (Mr. Randall's section), considerably lower down on the slopes of the hills capped by the Garland conglomerate, No. X. It is in fact one of the Oil Sands of Venango county; whereas the Garland conglomerate is the Second Mountain Sand. This is another proof that the oil sands struck by Mr. Beatty far below river level at Warren, must lie towards the *bottom* of the Chemung, if they be not the top layers of the Portage.*

---

* The above chapters are printed as an appendix to Mr. Carll's report of progress in 1874, not only for the use of the readers of that report, and to obey the organic law of the Survey requiring the publication of all data touching Pennsylvania Geology from whatever source, but also and chiefly for the instruction and guidance of that part of the Corps of the Survey to which is assigned the examination of the northern and north-western counties of the State. It will serve as a text-book, or guide in further explorations.

The Report of Progress in 1875, will commence with Chapter XV, and continue the description of the same series of Formations through M'Kean, Warren, Venango, Mercer, Crawford and Erie counties in Pennsylvania.

INDEX

TO MR. CARLL'S REPORT OF 1874.

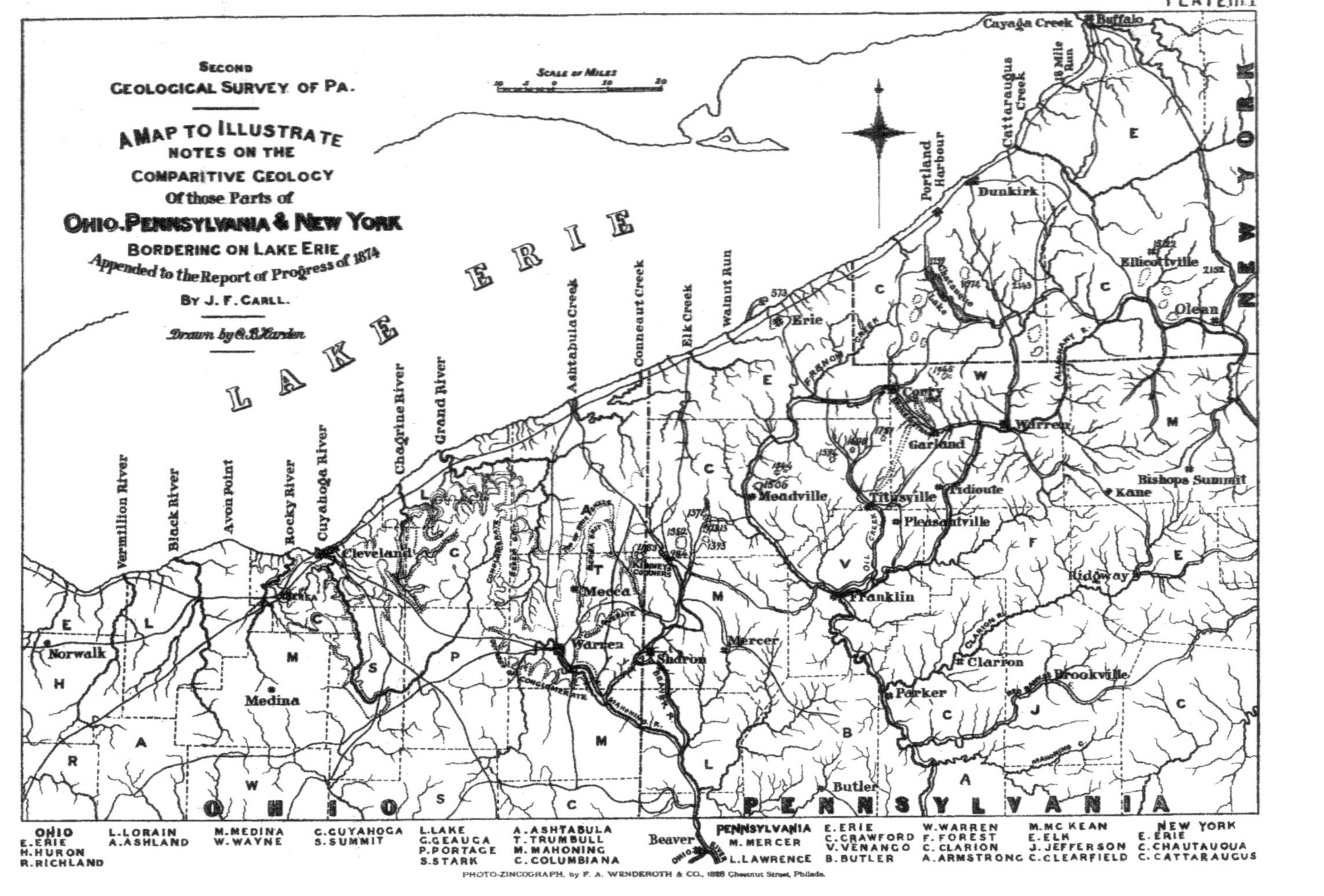

SECOND
GEOLOGICAL SURVEY OF PA.
A MAP TO ILLUSTRATE
NOTES ON THE
COMPARITIVE GEOLOGY
Of those Parts of
OHIO, PENNSYLVANIA & NEW YORK
BORDERING ON LAKE ERIE
Appended to the Report of Progress of 1874
BY J. F. CARLL.
Drawn by O. B. Harden
SCALE OF MILES
NEW YORK
PENNSYLVANIA
OHIO
LAKE ERIE
Cayaga Creek
Buffalo
18 Mile Run
Cattaraugus Creek
Portland Harbour
Dunkirk
Ellicottville
Olean
Erie
Corry
Warren
Garland
Meadville
Titusville
Tidioute
Pleasantville
Kane
Bishops Summit
Franklin
Ridgway
Vermillion River
Black River
Avon Point
Rocky River
Cuyahoga River
Chagrine River
Grand River
Ashtabula Creek
Conneaut Creek
Elk Creek
Walnut Run
Cleveland
Mecca
Warren
Sharon
Mercer
Clarion
Parker
Brookville
Butler
Norwalk
Medina
Beaver
OHIO
E. ERIE
H. HURON
R. RICHLAND
L. LORAIN
A. ASHLAND
M. MEDINA
W. WAYNE
C. CUYAHOGA
S. SUMMIT
L. LAKE
G. GEAUGA
P. PORTAGE
S. STARK
A. ASHTABULA
T. TRUMBULL
M. MAHONING
C. COLUMBIANA
PENNSYLVANIA
M. MERCER
L. LAWRENCE
E. ERIE
C. CRAWFORD
V. VENANGO
B. BUTLER
W. WARREN
F. FOREST
C. CLARION
A. ARMSTRONG
M. MC KEAN
E. ELK
J. JEFFERSON
C. CLEARFIELD
NEW YORK
E. ERIE
C. CHAUTAUQUA
C. CATTARAUGUS
PHOTO-ZINCOGRAPH, by F. A. WENDEROTH & CO., 1825 Chestnut Street, Philada.